AF846645

# Thinking in Physics

***Grenoble Sciences***

The aim of Grenoble Sciences is twofold:

– to produce works corresponding to a clearly defined project, without the constraints of trends nor curriculum,
– to ensure the utmost scientific and pedagogic quality of the selected works: each project is selected by Grenoble Sciences with the help of anonymous referees. In order to optimize the work, the authors interact for a year (on average) with the members of a reading committee, whose names figure in the front pages of the work, which is then co-published with the most suitable publishing partner.

Contact: Tel.: (33) 4 76 51 46 95
E-mail: grenoble.sciences@ujf-grenoble.fr
Website: https://grenoble-sciences.ujf-grenoble.fr

***Scientific Director of Grenoble Sciences***

Jean BORNAREL, Emeritus Professor at the Joseph Fourier University, France

***Grenoble Sciences*** is a department of the Joseph Fourier University supported by the ***French National Ministry for Higher Education and Research*** and the ***Rhône-Alpes Region***.

***Thinking in Physics*** is an improved version of the original book
***En physique, pour comprendre***
by Laurence VIENNOT
EDP Sciences, Grenoble Sciences'collection, 2011, ISBN 978-2-7598 -0656-0.

The Reading Committee of the French version included the following members:

– **Guy AUBERT**, Emeritus Professor, Joseph Fourier University, Grenoble 1. Scientific adviser CEA/DSM/Irfu
– **Jon OGBORN**, Emeritus Professor, Institute of Education, University of London
– **Jacques RICARD**, Emeritus Professor, Paris Diderot University, Paris 7. Member of the Académie des sciences
– **Madeleine VEYSSIÉ**, Honorary Professor, Pierre & Marie Curie University, Paris 6

*Translation from original French version performed by*
Chris COLLISTER, Jonathan UPJOHN and Nicole SAUVAL
*Typesetting*: Anne-Laure PASSAVANT
*Figures*: Simone GERLIER, Pixel project
*Cover illustration*: Alice GIRAUD, with extracts from: slits – W. KAMINSKI; water jets – Gorazd PLANINŠIČ; Doppler effect – Pbroks 13, Wikimedia; hot-air balloons – Jean-Simon ASSELIN, Flickr

Laurence Viennot

# Thinking in Physics

## The pleasure of reasoning and understanding

Springer

Laurence Viennot
Laboratoire de Didactique André Revuz
PRES Sorbonne Paris Cité,
Université Paris Diderot
75205 Paris Cedex 13

Translated from "En Physique pour Comprendre, Laurence Viennot, Collection Grenoble Sciences. Paris: EDP Sciences 2011"

ISBN 978-94-017-8665-2 ISBN 978-94-017-8666-9 (eBook)
DOI 10.1007/978-94-017-8666-9
Springer Dordrecht Heidelberg London New York

Library of Congress Control Number: 2014932679

© Springer Science+Business Media Dordrecht 2014
No part of this work may be reproduced, stored in a retrieval system, or transmitted in any form or by any means, electronic, mechanical, photocopying, microfilming, recording or otherwise, without written permission from the Publisher, with the exception of any material supplied specifically for the purpose of being entered and executed on a computer system, for exclusive use by the purchaser of the work.

Printed on acid-free paper

Springer is part of Springer Science+Business Media (www.springer.com)

# FOREWORD

Laurence VIENNOT's book is simultaneously both modest and very ambitious; both very practical and highly visionary. It is modest and practical in choosing to address the mundane everyday issues that confront every physics teacher every day, and to offer tried and tested ways of dealing with them, which any teacher could readily adopt. But it is ambitious and visionary in the very fundamental problems and issues it chooses to confront – issues that are all too often neglected or ignored, yet which can make all the difference to the experience of learning physics.

"*Thank you – you made me think*". This is one of the most telling, yet typical, responses of students to the exercises Laurence VIENNOT has devised for them. It is this attitude to what physics is–as involving hard but rewarding thinking–which has informed the work carried out by Laurence VIENNOT and her colleagues over several decades. The key to all this work is the constant search for the possibility of intellectual satisfaction for students. Physics, of course, offers many pleasures, but this surely is one of the most important for those who engage seriously with the subject.

In the best possible sense then, this book is seriously unfashionable. It does not go for the easy solutions, of "fun" with physics, which are often offered to try to make physics more enjoyable. Nor is it a book full of idealised exhortations, but rather one full of practical well-worked-out suggestions for ways to get students thinking hard, productively and pleasurably. As a teacher, you can expect it to challenge some of the things you habitually do and think, even to annoy you from time to time as it brings you up short against some common confusion or contradiction, including several that appear in many a textbook. See, for example, whether you really understand what lifts a hot-air balloon, or whether you too have without knowing it been sold a self-contradictory explanation.

For Laurence VIENNOT, the pleasure of physics lies in its great sweep and power, in the way a small number of principles govern a huge range of phenomena, with elegance, parsimony and internal consistency. If you share anything of this feeling for the subject, and want to know how to get students–real everyday students in any classroom–to share it too, then this book deserves your careful attention. It will give you a magnificent supply of insights and ideas, all of which you can put to use no matter what programme of physics you happen to teach.

Jon OGBORN
*Emeritus Professor of Science Education*
*Institute of Education*
*University of London*

# Foreword to the French edition

Just over ten years ago, in my foreword to one of the many books Laurence VIENNOT has written dedicated to the teaching of physics and aptly entitled *Enseigner la physique* [*Teaching physics* for the English version], I pointed out just how disaffected students had become with scientific pursuits, not only in the United States, but also in Europe, and France in particular.

As becomes clear from the very first sentence, this new book, *Thinking in Physics*, tackles the problem from an original perspective: "*Above all, Physics has got to be enthralling. That is what we all go on repeating, but here we are at the start of the third millennium faced with falling numbers, and the likelihood of a dearth of physicists in the near future*".

Don't misunderstand me, Laurence VIENNOT is not rallying to the flag of those educationalists intent on transforming all forms of teaching into some sort of play activity. Naturally, it is quite possible to learn while playing but her aim is altogether more ambitious, since what she is concerned with is not so much learning as with understanding; this involves a far more deeply motivating intellectual satisfaction. For many students or school pupils, physics doesn't go much beyond "*applying the formula*" to "*solve*" a problem to satisfy the conventions set out in the detailed instructions of the official programmes. The system is so strictly constrained and codified that students can learn without understanding; this is hardly very attractive.

*Thinking in Physics* is not what you would call an easy book, but happily the reader can be inspired (or intrigued) by dipping at will into the Contents. It's also a book which may puzzle and provoke, giving even the most well-informed of physicists something to think about. To give just one among the examples illustrating the cover, there is a great deal that needs to be understood before embarking on a flight in a hot-air balloon: see the "*instructional hot-air balloon*" of Chapter 6!

The layout of the website associated with the book (https://grenoble-sciences.ujf-grenoble.fr/pap-ebook/viennot) provides easy access, so that the reader can fully benefit from the numerous bibliographic references. Another of the examples on the cover illustration is that of the pierced bottles of Appendix D: here we have a marvellous accumulation of historical errors, and imagine my surprise, and my disappointment on discovering that the genius Leonardo DA VINCI himself had failed to verify his conclusions experimentally, despite it being so much easier than for his flying machines.

I recommend this book not only to all my colleagues engaged in teaching physics and other scientific disciplines, at whatever level, but also to students, future teachers and all those who take pleasure in understanding...

Guy AUBERT
*Emeritus Professor, Université Joseph FOURIER, Grenoble*
*Honorary Government Adviser (Conseiller d'État)*
*Former director general of the Centre National de la Recherche Scientifique*

# ACKNOWLEDGEMENTS

I am very much indebted to Philippe COLIN, Jean-Luc LEROY-BURY, Ivan FELLER and Stephanie MATHÉ for their friendly, intense and precious contribution to the discussions and reflections presented here. Furthermore, I thank them very much for kindly accepting to check how some research investigation results we performed together are quoted in this book.

# PREFACE

Above all, Physics has got to be enthralling. That is what we all go on repeating, but here we are at the start of the third millennium faced with falling numbers, and the likelihood of a dearth of physicists in the near future. Many teachers will reply that in any case, as it is part of the curriculum, physics must be engaging. But the question remains, how can it be done?

This book is an attempt to find a partial response. Partial, but focusing on one particular aspect which is essential for a great many teachers. Others might not see things in the same way since, naturally, educational objectives come down fundamentally to a political choice. But choice entails knowing what's available and is realistically attainable. The important question here is that of the intellectual satisfaction of the learners.[1] The reader can judge for himself the importance of helping the students experience the pleasure of the thought process in physics; this is a point to which we will come back to in the conclusion. In the meantime, what concerns us is to propose elements which could help those teachers who wish to go deeper into this topic.

Of course, teachers count for a lot more than the official texts, despite strong influence of the latter and, as far as giving intellectual satisfaction is concerned, many will say that they haven't waited for official encouragement in order to pursue these high-minded ideals. But precisely, because there are so many constraints, it's important to be realistic, and a degree of humility is required. However, there is every hope that the resources used to achieve a given beneficial effect can be widely shared within the realistic contexts of education or scientific information.

To what extent is it possible to teach under realistic conditions while at the same time encouraging the intellectual satisfaction of our students? And how?

The purpose of the book is to cover this question, albeit, of course, within certain limits. We shall focus our attention on how the learner can acquire the power of

1 Several researchers have noted the expression "intellectual satisfaction" in their research. VIENNOT L. (2006) Teaching rituals and students' intellectual satisfaction, *Phys. Educ.* **41**, 400-408 (http://stacks.iop.org/0031-9120/41/400); FELLER I. (2008) *Usage scolaire de documents d'origine non scolaire en sciences physiques. Eléments pour un état des lieux et étude d'impact d'un accompagnement ciblé en classe de seconde*, Thesis, Université Paris Diderot (http://tel.archives-ouvertes.fr/tel-00366318/); FELLER I., COLIN P. & VIENNOT L. (2009) Critical analysis of popularisation documents in the physics classroom. An action-research in grade 10, *PEC*, 17 **(17)** 72-96; MATHÉ S. & VIENNOT L. (2009) Concern for coherence in journalists and science mediators-to-be: are they open to this prospect? *PEC,* 11 **(11)** 104-128.

reasoning in physics. We do not speak, as such, of "scientific procedure(s)", or even somewhat more modestly of an "inquiry based approach". Essentially, our interest here lies in the intelligibility and consistency of thought formulated by a student, an author, a supposed authority or any other participant. Intelligibility and consistency: how do we persuade teachers and students alike not to give up on these two areas? How can we measure the satisfaction of students seeing themselves grow intellectually and rewarded by persistence?

These questions seem to assign the (supposed) beneficiaries of this process to a relatively passive role, as mere recipients of the message. Are we perhaps going in completely the wrong direction at a time when so much value is placed on individual activity?

Well, yes, and no. What we want to emphasize here is the importance of acquisition of understanding, in the strongest, active sense of the words. Scientific training is not just training someone to make discoveries, still less training directly in making discoveries. The future citizen, even the future researcher, will need to grasp what others have said about science. How we react to the results of others or the positions they adopt is crucial. Not even the research worker spends all his time alone in research, far from it, in fact. He also has an eye on what others are doing (or he should be), and wonders what to make of their work. "Having the solution to a problem is not sufficient, you need to know how to make use of it":[2] this maxim that we try to pass on to our students holds true in a much broader sense.

In invoking this ability, we refer to what is often known as the critical faculties. We are essentially interested here in one of its components: the pursuit of intelligibility, with the aim of assessing the consistency and field of application of the idea being analysed. This is therefore a somewhat restricted approach, since no attempt is made to examine issues such as the social relevance of the questions raised or the perceived image of the development of science in a given text.

However, neither is this a case of a process restricted to normally recommended practices (which most certainly have their use) of "checking the result" of a calculation via standard techniques such as dimensional analysis or even limiting cases.[3]

This book has a wider ambition, that of the intellectual satisfaction of our students gained through deeper understanding. In order to attain this, we rely on the great

2 Viennot L. (1997) *Corrigés modes d'emploi*, document LDAR (formerly LDSP), Université Paris Diderot, p. 2.

3 See below, Chapter 1, note 21. In this regard, it seems regrettable that this ability has been very little assessed in the French baccalaureat. See M. Rigaut thesis (2005) *L'épreuve écrite de physique au baccalauréat : analyse du point de vue du contrat didactique, une étude centrée sur les années 1999 et 2000*, Université Paris 7 (www.matthieurigaut.net/public/docs/these_didactique_matthieu_rigaut.pdf).

internal consistency and predictive power of the physical theories currently in use.[4] The question is how to get students to be aware of this.

However, before waxing too lyrical about the virtues of intellectual satisfaction, it has to be acknowledged that anyone seeking to grasp explanatory reasoning will encounter the prefix "satis" in the word "satisfaction": so what will "suffice" for gaining an understanding? OGBORN *et al.* (1996)[5] rightly emphasise that reasoning in the physical sciences in terms of explanatory power is rather like the tip of the iceberg: underneath is a mass of implicit theory underpinning both the explanation provided and the question posed.[6] The authors propose this analysis in relation to teachers' explanations in class, though we note in passing that it applies also to so-called common-sense reasoning: in both cases, the argument relies on elements accepted without discussion (below the iceberg's waterline), either because they were previously learnt in class (and already partly digested) or because self-evidently attributed to common knowledge (BACHELARD).[7]

Pursuing this analogy, it seems that an essential basis for the intellectual satisfaction of anyone seeking understanding is his own adaptability to the waterline of the iceberg with which he is confronted. We are all familiar with the insecurity created when this is not the case, and any self-respecting teacher will take pains to avoid such a situation. Incidentally, it so happens that certain feelings of security may well be inappropriate: we shall see cases where what rouses the student's interest is the sudden realisation that something previously thought to be understood, actually hasn't been. In particular, (always, in fact) we have to be content with only a certain comprehension of the concepts involved. "Satis", sufficiently, where is this optimal waterline, enabling greater comprehension to emerge without too many pointless or risky explanations? Naturally, in this book there is no absolute answer, simply explorations around the accepted norm.

Of course, we also feel that the acquisition of a line of reasoning is not merely "sufficient" acceptance of each of the links in the chain of the argument. Contrary to this intellectual reductionism, it is possible to experience flashes of insight that are the result of an unspecified mixture of intuition and rationality. According to DE BROGLIE,[8] "A theory that succeeds, at a single stroke, in achieving a vast synthe-

4 For example: NEWTON's laws. On the subject of excessive relativism, see OGBORN J. (1997) *Constructivist metaphors of learning science*, Science & Education, **6**, 121-133.

5 OGBORN J., KRESS G., MARTINS I. & MCGILLICUDDY K. (1996) *Explaining Science in the Classroom*, Buckingham: Open University Press, p. 13.

6 They illustrate this with the behaviour of lithium, sodium or potassium when a piece is thrown into water, and the explanation linked with the periodic table of the elements, or even comments in molecular terms on the changes of state in the water.

7 BACHELARD G. (2002) *The Formation of the Scientific Mind* (Translation Mary MCALLESTER-JONES), Manchester: Clynamen Press.

8 DE BROGLIE L. (1941) *Continu et discontinu en physique quantique*, Albin Michel, Paris, p. 87.

sis (...) undeniably appears to the theorist as a thing of beauty, and might persuade him that it does indeed encompass a good part of the truth."

Can we not imagine that this kind of feeling may be encountered at all levels of the learning process, thus encouraging students to delve still further into intellectual inquiry?

While not going as far as DE BROGLIE's "vast synthesis", it is nevertheless often the case that intellectual pleasure is associated with stretching the possibilities of the thought process. Hence, the convergence of conclusions derived from thought processes taking different paths can be a cause for intellectual satisfaction. This is far removed from the idea that, in terms of pleasure, just a single explanation can suffice. Whether it's one explanation for a vast field of phenomena, or several paths leading to a single conclusion,[9] the intellectual pleasure can be immense, even though ultimately the explanation which seems most "economical" will be described as the most "beautiful".[10] A single demonstration is "sufficient" to convince us that a conclusion is valid, but in terms of pleasure, who can claim that any one explanation is enough?

With some justification, it has often been said that one learns better when well motivated, and that a prime motivating factor is the practical applications of science and the way it ties in with the technical and functional demands of society. Likewise, images, for example, of cosmology and nanoscale phenomena have the potential to stimulate and fire the imagination. In a somewhat different vein, the framework in which activities are carried out can be a galvanizing factor; for instance, one might mention the university projects or the *Travaux Personnels Encadrés* (supervised individual work) launched in France in the *classe de Première* (Year 12) in 1997. However, although the importance played by such factors in the desire to acquire or develop critical faculties cannot be denied, in the present discussion they have only secondary roles. For here indeed is the central question in this book: in accompanying a student on the road to reason, can we provide her/him with intellectual pleasure independent of the usual motivational factors? Surely, the effort, and indeed it is an effort, is at least worth a try?[11] The proposed paths will necessarily involve abstraction, but at a level that should always be accessible. The associated pleasure can be compared rather to a hike in the mountains (a sport anyone can do) than to sunbathing on the beach.

---

9 See the example of the hot-air balloon in Chapter 6.

10 R. FEYNMANN wrote *The character of physical laws* (1965): "But the most impressive fact is that gravity is simple (…). It is simple and therefore it is beautiful".

11 The idea of intellectual effort is very often excluded from the definition of what scientific popularisation ought to be. "Popularisation boasts of offering science without pain", if B. JURDANT (1975) is to be believed, La vulgarisation scientifique, *La Recherche,* **53**, 149.

Of course, the previous comments echo those that often recur in our official teaching texts, and in that sense, they may not seem terribly original. However, the whole question is one of shifting the focus of what we do in practice to make the most of this official encouragement. Are these exhortations pointless in the real world? We often hear it said that students do not have sufficient critical faculties and that teachers do not have the time to help them with this. For our part, we refuse to equate realism with fatalism. Even so, this book does not attempt to deal with all the practical methods for promoting the learning of physics, especially as far as experimentation is concerned. It is to be taken simply as an appeal for greater shared pleasure, seeing beyond some of our conventional teaching rituals.

The suggestions which follow are illustrated by examples of appropriate teaching practices, typically at the end of secondary level or at the start of university, but applicable in principle to other levels. The content of the associated physics is deliberately simple. Stumbling blocks related to common tendencies in thinking and the rigid nature of teaching practices are highlighted with the aim of opening up choices and enabling better-informed teaching decisions. In particular, the ingredients commonly employed to make physics more "attractive" are reconsidered, either in the form of rather surprising and unconventional experimental mini-demonstrations, or in texts taken from popular science.

Thus, the Part I of the book analyses and illustrates one or two components of a fruitful thought process in physics, and high on the list of tools and techniques mentioned are functional dependencies and associated graphs, whose illustration is deliberately restricted to a few simple cases with additional practical suggestions. The Part II builds on the benefits of addressing different phenomena within the same formal approach, and different approaches used to analyse the same situation. The Part III puts into perspective the relative merits of experiments conducted along the lines of "mini-demos", when they are so amazing that they dazzle the critical faculties. Normal use is completed with an in-depth reflection which, in particular, takes account of some tendencies commonly seen among teachers. The topics covered then finally merge with popularisation.

In the conclusion we return once more to the idea that, both as teachers and popularisers, there are more choices available to us than might appear to be so at first sight, and hence that our responsibilities must be taken all the more seriously.

Laurence VIENNOT

*September, 2013*

# Contents

# Part I

## LEARNING TO THINK: WORDS, IMAGES AND FUNCTIONS

# Chapter 1

# Essential tools for comprehension

## 1.1 Words we have to understand

The thinking process which is our starting point, is fraught with dangers. If we say that a student has to understand the terms used for him to grasp the topic, we end up with a "Mission: Impossible". Understanding? Well, yes, but up to what point? The precise level of comprehension aimed for would have to be defined each time, or at least a level giving a (minimum?) intelligibility for a given reasoning process. Just to make this question of "pseudo-evidence" clearer, here are a few examples worth looking at a bit more closely:

- A mechanics text talks about a "particle", or a "material point". We might consider a point of mass $m$ to be rather too limited, and not exactly physical given that this seems to imply an infinite density. In getting to grips with this model, we could be tempted to take this a bit further.
- One exercise in optics is to find the "region of space" visible in a mirror by an "eye" at a given distance from this mirror that is the square of some particular side. A line of reasoning in response to such a question should assume that a "region of space" (described by length, area, angle, or solid angle?) and "eye" (modelled by a point, a pupil in series with a lens and a screen, a visual system including the corresponding area in the brain?)[12] have some meaning.
- The fact that a motion may be "uniform" is independent of its direction, but to say that a vector field (a magnetic field, for example) merits this attribute is to imply that it is uni-directional: an essential piece of information in understanding the reasoning processes which apply in this area.

When it comes to helping a learner acquire a line of reasoning, the difficulty for the participant is one of being just at the boundary between the evidence where the explanation might be tedious and that of an unsuspected incomprehensibility: if a Chinese term is put in, then there is clearly a question of meaning, but frequently,

12 See Chapter 4, Putting things into practice, exercise 4.2.

there is nothing to draw attention to a term whose familiarity obscures the fact that its meaning may not be obvious. This question of the meaning of the terms used is crucial in the popularisation of science, though it is by no means the only obstacle. Hence the use of metaphor, which is widespread within some disciplines, may well lead to pitfalls: do black holes and dark matter represent the same kind of "black darkness"? However, staying on relatively familiar, school-oriented territory, this problem posed by word meaning may be more serious than imagined. What are we to say about "thermal agitation", omnipresent in thermodynamics courses after Year 11? Do we then talk about speed, or kinetic energy?[13]

Explaining the meaning of the words is typically an area where it is only by the use of examples that interest can be maintained.

Obstacles to rational thought are sometimes more subtle however, lurking in the silences of the unsaid, things neither made explicit nor evoked by an image. This is without doubt the most frequent category in our examples. However, just as for vocabulary, the register is one of measure and nuance: naturally, there is no question of spelling everything out. Choices must be made depending on the risk of not being understood, and hence assessed according to the objective.

Between evidence, redundancy and underestimating the difficulties, the margin is sometimes narrow and the explanatory route hard to negotiate. There are no general rules here, but rather a few examples to contrast the offhand manner sometimes to be found in teaching and popularisation with the benefits that can be obtained at a reasonable cost thanks to just a little effort.

## 1.2 The image; it is really doing its job?

*In collaboration with Philippe Colin*

In order to conclude the verbal explanation and/or give credence to a fragmentary description, images may be used. Clearly, as a powerful adjunct to discussion, the image scarcely needs defending since it is rarely questioned by teachers. However, especially if we are to believe the findings of research into the use of illustrations in

13 See Chapter 3, notes 48 and 49. On this topic of the meaning of words, Gunstone R. & Watts R. wrote in 1985 (Force and Motion. In Driver R., Guesne E. & Tiberghien A., *Children's ideas in science*, Milton Keynes: Open University press): "Language which is meaningful to teachers may, because of students' views of the world, have a quite different (even conflicting) meaning for students. If we are not sensitive to this, we can unwittingly reinforce the very views we want to change." Cited by Klaassen K. & Linjse P. (2010) *Interpreting students' discourse and teachers' discourse in science classes: an underestimated problem?* In Designing Theory Based Teaching-Learning sequences for science Education. Kortland K. & Klaassen K. (Eds.), CDβ Press, Utrecht.

teaching manuals or exam papers, a certain number of doubts may be expressed:[14] Figures 1.1 to 1.6 provide a few examples taken from the field of optics which might be found in teaching, both at an elementary level and with more advanced students. These examples of images do not include errors *per se*, but are wholly consistent with an interpretation which poses problems from a physical point of view (ref. in note 14). On the left of each figure are diagrams or figures encountered in school or university textbooks; these are put into context. On the right, are interpretations formulated by students, and where necessary, the caption which can create the problems, followed by a suggestion of ours or an idea to work on.

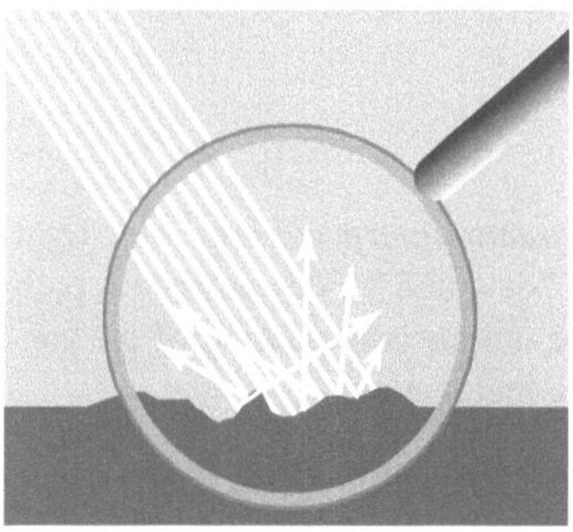

***Typical student interpretations***: The "rays" might be visible...

- The magnifying glass is used to show that the rays are very small (Year 11).
- The purpose of the magnifying glass is to see the path of the rays, which are so infinitesimally small that we cannot see them (Year 13).
- The figure shows that light travels in a straight line (thanks to white light, which we can see) (Year 13).

***Ideas to work on:*** A beam of light illuminates the rough surface, and this light is scattered in all directions. The "rays" are theoretical constructs and not visible objects.

***Figure 1.1*** *- A model for a rough surface which scatters light.*[15]

14 Colin P., Chauvet F. & Viennot L. (2002) Reading images in optics: students' difficulties and teachers' views, *International Journal of Science Education*, **24** (3) 313-332; Viennot L., Chauvet F., Colin P. & Rebmann G. (2005) Designing Strategies and Tools for Teacher Training, the Role of Critical Details. Examples in Optics, *Science Education*, **89** (1) 13-27. Regarding interference and diffraction, see the thesis of P. Colin: Colin P. & Viennot L. (2001) Using two models in optics: Students, difficulties and suggestions for teaching, *Physics Education Research, American Journal of Physics Sup.* **69** (7) S36-S44. For a summary of these topics: Viennot L. (2003) *Teaching physics*, Kluwer Ac. Pub., Dordrecht, Chapters 1 and 5. See also: Ambrose B.S., Shaffer P.S., Steinberg R.N., McDermott L.C. (1999) An investigation of student understanding of single-slit diffraction and double-slit interference, *American Journal of Physics* **67** 146-155; Wosilait K., Heron P.R.L., Shaffer P.S., McDermott L.C. (1999) Addressing student difficulties in applying a wave model to the interference and diffraction of light, *American Journal of Physics* **67** S5-15; Vokos S., Shaffer P.S., Ambrose B.S., McDermott L.C. (2000) Student understanding of the wave nature of matter: Diffraction and interference of particles, *American Journal of Physics* **68** S42-S51. More generally, the "grammar of the image" and ways of decoding it appear in numerous studies. For example, Kress G. & van Leeuwen T. (1996) *Reading Images: the Grammar of Visual design*, London: Routledge & Kegan Paul.

15 Karplus R. (1969) *Introductory physics. A model approach*, Benjamin INC., New York, W.A., p. 124. Illustrations showing white "rays" on a grey background are to be found in numerous books, e.g. Crawford F.S. (1965) *Berkeley Physics Course*, Waves, McGraw-Hill Book Company, New York; however, here the effect of realism is accentuated by the magnifying glass.

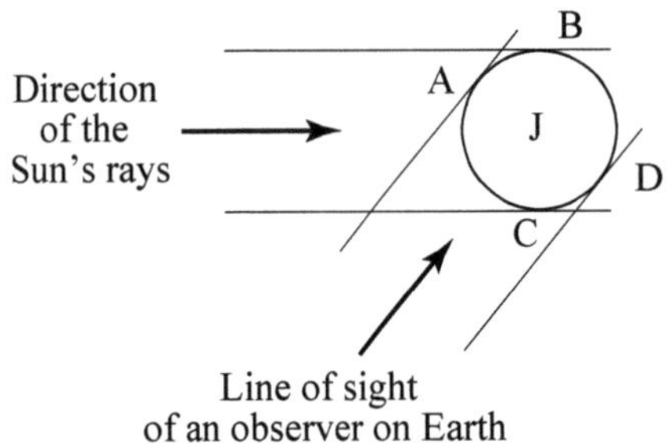


***Typical student interpretations:*** The slanting arrow seems to be interpreted as a ray (visual?):

- The rays from Earth cross the rays from the Sun (Year 11).
- CD and AB are both in shade (Year 11).

***Suggestion:*** Avoid using the same symbols for rays and line of sight.

***Figure 1.2*** *- This figure is intended to explain the visibility of Io, one of Jupiter's satellites.*[16]

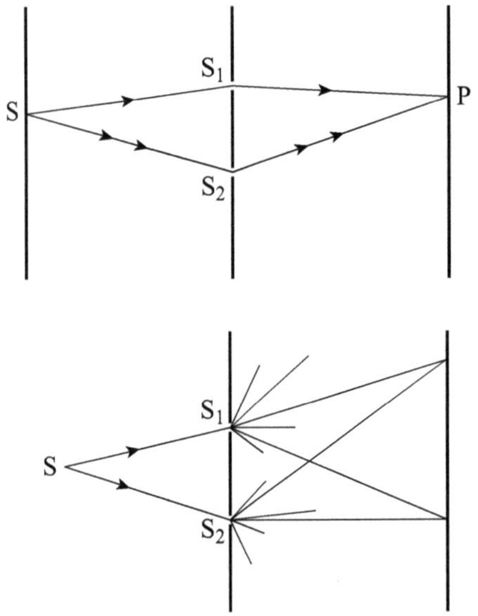


***Typical student interpretation:*** The light may have "deviated": Each emerging path may be the unique continuation of the corresponding incident ray.

- The light has deviated (Year 11).
- The light has deviated (blanked out). The light cannot follow these paths (Year 13).

***Ideas to work on:*** What happens at the holes is a diffraction phenomenon.

***Suggestion for a less ambiguous drawing:*** This diagram suggests the phenomenon of diffraction, and shows how to analyse the illumination of the screen at various points. This involves selecting the relevant light paths for each point on the screen (backward selection).

***Figure 1.3*** *- Principle of interference using* YOUNG*'s holes*
*(S: point source of light,* $S_1$, $S_2$*: holes in the first screen, P: detector).*

16 Very simplified version of an image shown in BOTINELLI L., BRAHIC A., GOUGUENHEIM L., RIPERT J. & SERT J. (1993) *La Terre et l'Univers*, Hachette, Paris, p. 121.

***Typical student interpretation:*** Point M could be the image of a point at infinity, the rays may have "deviated"...:

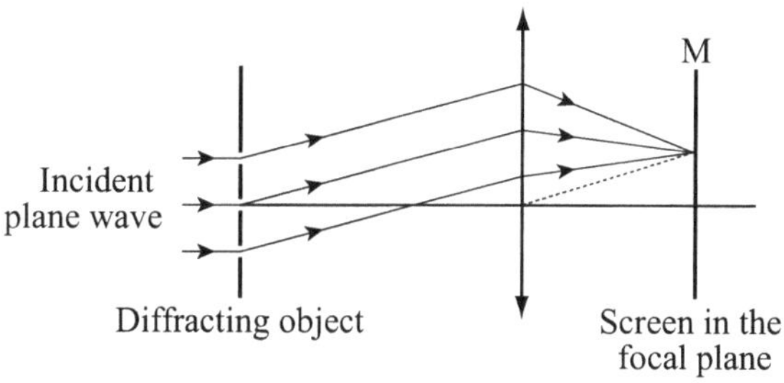


– All the pencils of light from the three slits converge to the same point M on the screen. These three beams originate at infinity (...) (University first year)

– The incident rays are parallel to the axis and deviate by the same angle. All emergent rays arrive at the lens parallel to one another. (University first year)

– The beams arriving at the slits are mutually parallel. After passing through the slits, the beams are deflected but stay mutually parallel. (University first year)

***Ideas to work on:*** The light paths involved do not correspond to one and the same plane wave. A wave model is appropriate beyond the slits, and there is generally no phase coincidence at M.

***Figure 1.4*** *- Obtaining the diffraction pattern of an object.*

***Comment on this figure (problematic):*** For a given direction $\vec{u}$ the wave surfaces corresponding to the diffracted rays are planes perpendicular to $\vec{u}$. (school textbook)

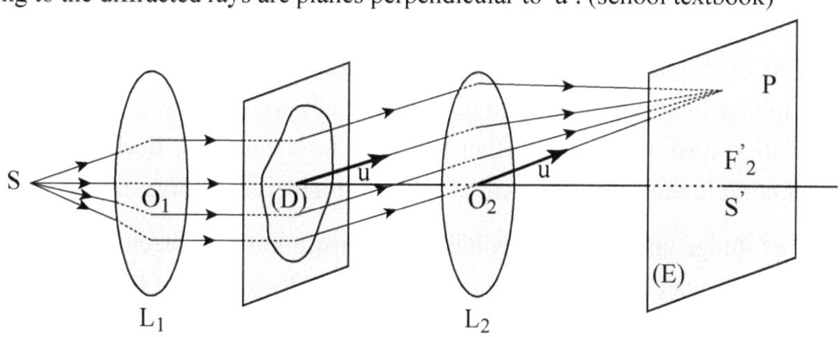


***Ideas to work on:*** It is inconsistent to talk of the wavefront or plane of the wave associated with diffracted rays; the lines represent light paths for which the phase in such a plane is not normally the same (whence standard calculations of the resultant light intensity). The fact that these paths are parallel goes too far in suggesting that this is a plane wave.

***Figure 1.5*** *- A diagram in a textbook.*[17]

17 Bertin M., Faroux J.P. & Renault J. (1986) *Optique et physique ondulatoire ; Optique géométrique et optique physique. Phénomènes de propagation,* Cours de physique, 3e édition, Marketing, Paris. See also, for example, Figures 10.10, (c and d), in Hecht E. (1987) *Optics*, 2nd edition, New York: Addison-Wesley, which suggest that, for a given direction, an object will diffract a plane wave (Fig. 10.10d: "These rays bundles correspond to plane waves which can be thought of as the three-dimensional Fourier components") whose amplitude depends on how the "rays" interfere.

Here again, there are no rules for qualifying, a priori, an image as useful or potentially dangerous. To decide this, it is essential to take into account the target population.

Figure 1.6(a) shows a photograph of the type often used to demonstrate rectilinear propagation, suggesting the existence of "actual" rays.

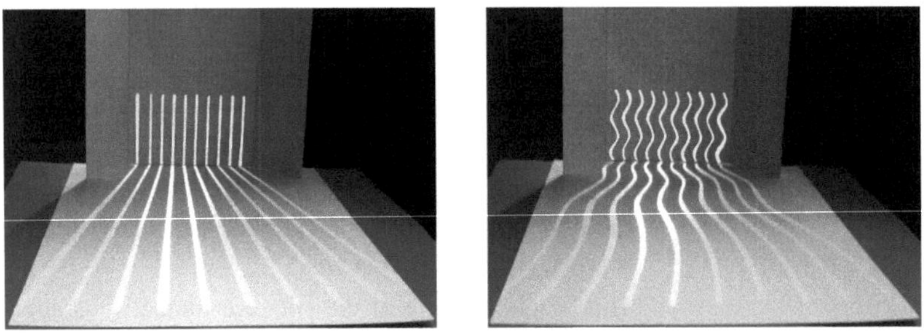

(a) (b)

***Figure 1.6*** - (a) *A small lamp behind a screen in which parallel slits have been cut, produces traces of light on the surface; arrangement* (b) *avoids oversimplification in this respect. In both cases what is seen is a set of shadows: The observed traces are not rays.*[18] *Photos W. Kaminski.*

In an introductory course on elementary optics, casual use of the photograph could well give the impression that visible light rays are passing over the horizontal surface, in nice straight lines, just as they should. In Figure 1.6(b), the wavy lines made by the light bring the arrangement back to the category they should be in, *i.e* that of shadows. Each point of a trace of light is visible as a result of scattering of the received light, which got there via a rectilinear path not parallel to the surface. This reinterpretation also resolves a further problem: how could the horizontal so-called "rays" not contain their source, located ten centimetres above the sheet?

This kind of image and the corresponding demonstration have been criticised on the grounds that they can lead to errors in understanding the nature of light. Light is invisible from the side, and we don't see it propagate the way a train goes past: it is only when light strikes a surface that it is detected. However, these criticisms should doubtless be reconsidered in those cases where the use of such "visualisation" is not associated with some conceptual goal on the nature of light itself.

18 Viennot L. (2004) The design of teaching sequences in physics - Can research inform practice? Lines of attention. Optics and solid friction In *Research on Physics Education, Proceedings of the International School of Physics Enrico Fermi (Italian Society of Physics), Course CLVI,* Societa Italiana di Fisica, Bologna, 505-520. See also Viennot L. (2006) Teaching rituals and students' intellectual satisfaction, *Physics Education*, **41**, 400-408 (http://stacks.iop.org/0031-9120/41/400). Concerning lenses: Viennot L. & Kaminski W. (2006) Can we evaluate the impact of a critical detail? The role of a type of diagram in understanding optical imaging, *International Journal of Science Education*, **28** (15) 1867-1895; Galili I. & Hazan A. (2000) Learners' knowledge in optics: interpretation, structure, and analysis', *International Journal of Science Education*, **22** (1) 57-88.

Another common example is that of "false colour" images at nanoscale or in astrophysics. Photos of atoms, making them look like pretty blue spheres won't, of course, mislead a physicist, but it would be better to warn the young readers of popular magazines that the atom is not actually blue.[19]

Comments made about images should be considered without losing sight of the overall context.

## 1.3 Graphs and functions

If reasonably well presented, graphs are generally quite successful at conveying the intended message, and hence treated with less caution than images with figurative or symbolic components. Sometimes, however, their lack of ambiguity, their forbidding abstraction may paradoxically make them appear threatening. As opposed to images, which are often credited with excessive transparency, graphs do not appear to be a natural aid in the reasoning process, but almost as a compulsory exercise, impenetrable to all but the brightest students. A graph linking two spatial variables, such as the functional representation $y = f(x)$ may be OK.[20] However, when it comes to a function involving time, such as $y = f(t)$, then perplexity rules.

Justice must be done to this theoretical tool in which a graph represents a function. In a constitutive sense, it simply links with the analysis of functional dependence, this being the cornerstone of reasoning in physics.

Let's come back to this expression "functional dependence". It goes beyond the idea that a formula offers the possibility of calculating something. From a formula we can calculate a variable using other known quantities. However, if the formula is now read as a functional dependence, this opens up the field of deductive possibilities. A whole set of transformations is now involved in the analysis:[21] starting from some arbitrary state, if we increase one quantity while holding another constant, then such and such a thing happens... Leaving behind the numerical evaluation of a state we pass on to the transformation of mutually dependent variables. If changing one

19 See for example in *La lumière et la matière* (MNSER 2005) a pamphlet written for secondary school pupils during world physics year, Figure 21, p. 13, in which the caption is limited to "A single xenon atom (light blue) on a nickel surface, seen by..."

20 Even in this case, a curve may be inappropriately understood as the path of a route on a geographical map. If in a different reference frame, the trajectory of a moving object is different, even though the path traced out is not altered by such a change. This distinction is not at all intuitive (VIENNOT L. (2001) *Reasoning in Physics - The part of common sense*, Dordrecht: Kluwer Ac. Pub., 52-55; original study by SALTIEL E. & MALGRANGE J.L. (1980) Spontaneous ways of reasoning in elementary kinematics, *European Journal of Physics*, vol. 1, 73-80).

21 In particular, this provides the liberty to explore what happens in those "limiting cases" in phenomena where a variable in the formula tends to zero or infinity. If these limiting cases are perhaps known via some other means (common sense, problem already solved, etc.), then this provides a check on the formula.

variable fails to change another, then an invariance has come to light. In other words, one analysed situation represents a wider set of situations. If the invariance is surprising (and we shall see examples of this) then this is naturally much more interesting than if it had seemed self-evident.

These functional considerations quickly lead to an area far beyond merely technical issues, and hence to the discovery of the descriptive power and compact elegance of physical theories. Apart from the merely practical, "toolkit" side of things, more personal, more affective aspects may be involved on such occasions, namely the satisfaction that comes from understanding, a sometimes profound, maybe ephemeral contentment but (who knows?) perhaps followed by more far-reaching effects than are immediately apparent.

# Chapter 2

# Some surprising invariances

## 2.1 Introduction

The colour of the experimenter's hair does not generally influence the outcome of his experiment, any more than would the positions of the planets or the value of some stock market indicator. So why, if there are so many of them, should we point out factors which have no influence over the phenomenon being studied?[22] When teaching an idea within the text of a standard exercise, it is normal practice to mention only that which is strictly "relevant", and, in most cases, identified with the variables whose symbols appear in the algebraic expressions used. This results in a draconian selective filtering process. The history of ideas is often one of abandoning parameters hitherto thought to be relevant. The typical wording of exercises reflects this acceptance and adds a layer of simplification for other parameters which should, in principle, be taken into account: how many times have we seen "ignoring friction...."? It's already complicated enough, we hear them say.

Temporarily leaving aside the question of a sometimes over-simplified model, it's worth reflecting on what we don't say on what doesn't count. Or rather, on what we don't emphasise, as such. The value of an invariance (or a non-dependence) lies in its surprising character.

## 2.2 The speed of light *in vacuo*

It is fairly well known that light in a vacuum, or in air, which is pretty much the same, travels very fast. Its speed, denoted **c**, is very close to 300,000 km/s ($\mathbf{c} = 2.99792458.10^8$ m.s$^{-1}$). A common symbol and a value: but have all aspects of

22 S. Rozier explored this question in 1983 (*L'implicite en physique : les étudiants et les fonctions de plusieurs variables*, mémoire de tutorat (year 5 at university) DEA de didactique, Université Paris Diderot). Viennot L. (1982) L'implicite en physique : les étudiants et les constantes, *European Journal of Physics*, **3**, 174-180; and Viennot L. (2001) *Reasoning in Physics - The part of common sense*, Dordrecht: Kluwer Ac. Pub., Chapter 9.

this question been considered? If we go into it a bit more deeply, we end up with considerations which are a lot more interesting than that.

The speed of light? What kind of radiation are we talking about? For visible light, everyone knows that this involves an infinite number of wavelengths (or frequencies, if you prefer) each corresponding to a colour. The same remains true outside the visible spectrum (radio waves for example), and we no longer simply talk about colour (though often keeping the adjective "monochromatic" for a given wavelength). Do all these waves travel at the same speed?

The second thing to bear in mind is that when referring to speed, we usually have to state the reference frame in which it is measured.

In fact, if we don't mention it, it is because it doesn't count. For all wavelengths (*i.e.* for all frequencies), electromagnetic radiation, visible or otherwise, has the same speed ("phase velocity" in scientific terms) in a vacuum and in all Galilean reference frames.[23] This last point is particularly startling, and it took some time to convince the world of it. Associated with this astonishing observation are, from 1881 onwards, the celebrated names of MICHELSON and MORLEY, but it was not until 1905 that EINSTEIN in his famous paper on *Special Relativity*[24] finally settled the question.

At the start of university, students are, for the most part, able to say that the speed of light is constant, and can give a value for **c**. However, many do not know, or at least cannot express, the invariances we have just reviewed,[25] and that is why it is worth emphasising them.

## 2.3 Propagation of mechanical signals

A mechanical signal such as a disturbance propagating on a string or a sound in the air can be described by a wave whose speed depends only on the medium.[26] Let us now

23 An infinite number of reference frames in mutually rectilinear uniform translation in which NEWTON's laws apply for objects moving much slower than the speed of light.

24 EINSTEIN A. (1905) Zur Elektrodynamik bewegter Körper, *Ann. d.Ph.*, **17**, 892-921 (translation by SOLOVINE, GAUTHIER-VILLARS, 1955, **5**). See also: EINSTEIN A. (1907) Über das Relativitätsprinzip und die aus demselben gezogenen Folgerungen, *Jahrbuch der Radioaktivität*, **4**, 411-462 & **5**, 98-99.

25 VIENNOT L. (2001) *Reasoning in Physics - The part of common sense*, Dordrecht: Kluwer Ac. Pub., 155-156.

26 This is in a "non-dispersive" medium in which waves of varying frequency propagate at the same speed, the form of the "bump" being preserved. The expression for the phase velocity (or celerity) of sound is $c = \left[\frac{RT\gamma}{M}\right]^{1/2}$, where $R$ is the ideal gas constant, $M$ is the molar mass of the gas, $T$ is the absolute temperature and $\gamma$ is a coefficient which depends on the number of atoms in the gas molecule and on $T$. The speed of propagation of a disturbance on a string is $c = \left[\frac{T}{\mu}\right]^{1/2}$ where $T$ is the tension and $\mu$ the mass per unit length of the string.

ask what happens if the source of the signal is more powerful–if the string is struck "harder" or someone shouts louder: a good proportion of people questioned[27] predict that the disturbance will propagate faster. And this, despite the fact that the commonly taught formulation (*i.e.* the speed depends only on the medium) is well known to them. What is certain is that they just have not fully realised what this means *i.e.* a surprising thing: that propagation speed is independent of the power of the source.

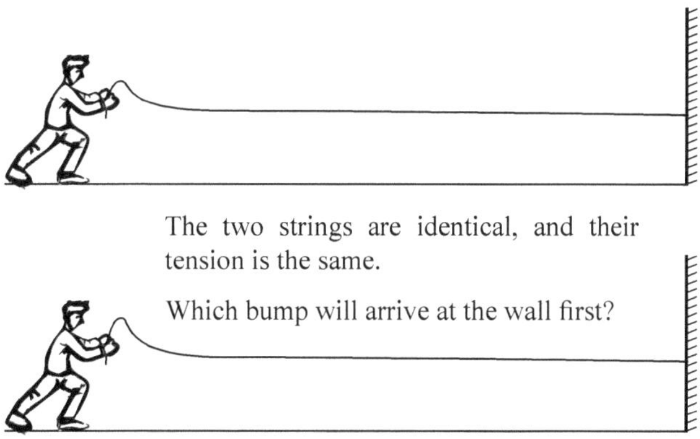


***Figure 2.1** - A situation which underlines the meaning of a well-known statement: "for a stretched string, the speed of propagation of a disturbance depends only on the mass per unit length and the tension". In this model, the race between the disturbances is a foregone conclusion: it's a dead heat!*

There is some benefit to be gained from such knowledge, *i.e.* the ritual statement that propagation speed depends only on the medium, since it implies that this surprising invariance will be questioned by some, or emphasised for others. It is surprises and unexpected phenomena of this kind which illustrate how physical theories are not just a familiar collection of analyses of situations we know how to handle; they have great unifying power for specific cases that we might have thought distinct, but which in fact, from certain points of view, prove to be otherwise.

Sometimes, the merest hint of an explanation is sufficient to induce these welcome surprises, these new and unexpected insights.

## 2.4 Coefficients of friction

The value of a normal component of contact force $F_N$ and that of its tangential component $F_T$ are coupled by static or dynamic coefficients of friction for two interac-

27 Maurines L. Spontaneous reasoning on the propagation of visible mechanical signals, *International Journal of Science Education*, 14 (**3**) 279-293; Viennot L. (2001) *Reasoning in Physics - The part of common sense*, Dordrecht: Kluwer Ac. Pub., 143-144. See also Wittmann M.C. (1998) *Making Sense of How Students Come to an Understanding of Physics: An Example from Mechanical Waves*, Unpublished Ph.D. dissertation, University of Maryland; Wittmann M.C., Redish E.F. & Steinberg R.N. (2003) Understanding and Addressing Student Reasoning about Sound, *International Journal of Science Education*, **25**: 8, 991-1013.

ting surfaces. In standard elementary courses on friction,[28] the coefficient $\mu_s$ enables the maximum allowable value of the tangential component without slipping to be calculated: $F_T \leq \mu_s F_N$ and the coefficient $\mu_d$ enables the value of the tangential component, once slipping has started, to be found: $F_T = \mu_d F_N$. For a rectangular block of mass $m$ sliding on an inclined plane (at an angle $\theta$ to the horizontal), the written notation of the balance of forces and the fundamental principle of dynamics (air friction being "negligible") gives the value for tangential acceleration (axis downwards): $a = g\,(sin\theta - \mu_d\,cos\theta)$.

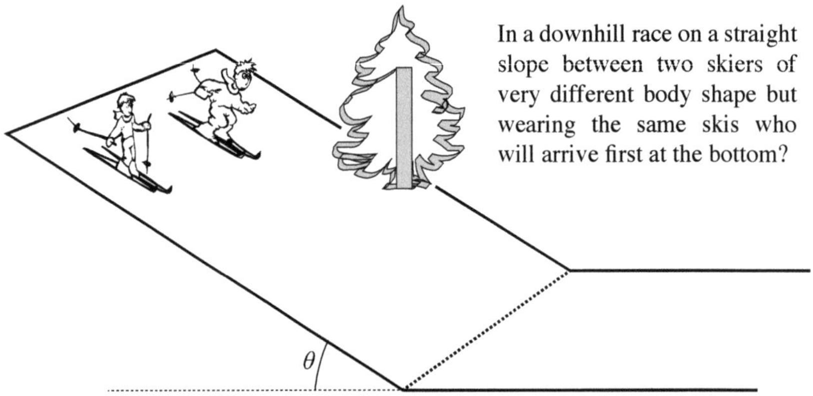


***Figure 2.2*** *- A situation which underlines the meaning of the standard solution to an exercise involving friction: "the acceleration of a skier along the path of steepest descent is: $a = g(sin\theta - \mu_d\,cos\theta)$", where $\mu_d$ is the coefficient of sliding friction and g is the acceleration due to gravity. According to this model, the race between the skiers is a foregone conclusion: dead heat. Why is this so improbable?*

Let us imagine (this might be a first year university exam question) that this block represents a model of a skier on the line of steepest descent. There is the solution, presented to a tutorial group.[29] What can we usefully add?

Here's a question: if two skiers, otherwise identical in every respect, have skis of different widths, does the solution above predict anything about their performance? This ski width does not appear in the expression for acceleration, $a$, but is it not relevant? Or else where is it hiding? In the coefficient $\mu_d$ perhaps? Otherwise we simply have to resign ourselves to the fact that the solution to this exercise predicts the simultaneous arrival of these two skiers on skis of different widths.

So we come to the question we could so easily have ignored: what exactly **do** these coefficients depend on, and what **don't** they depend on? It's surprising that $\mu_d$ and $\mu_s$

---

**28** For more information on the history and limitations of this simple model, read Besson U., Borghi L., De Ambrosis A. & Mascheretti P. (2007) How to teach friction: Experiments and models, *American Journal of Physics*, **75** (12) 1106-1113.

**29** See Chapter 4, exercise 4.4.

do not depend on the area of contact. We can discuss the value of this model, but at least we should realise what it means.

The model can be completed with the idea that, although in the case of a greater contact area there are more asperities, they are less crushed and hence play a less active role in friction. With the one cancelling out the other, the result could be the same. In any event, this invariance is more or less explicitly what is being taught today. While it may be surprising, the mere fact of emphasising it draws the students into intellectual activity far more gratifying than simply assigning a value to a coefficient to be mechanically plugged into a formula for calculating some other quantity.

## 2.5 When mass doesn't count

While the symbol for mass is found right from the outset in course books and exercises books in elementary mechanics, it is, astonishingly, among the variables which don't seem to count for very much. This is also puzzling in the light of common experience.

Quite simply, if the forces involved are proportional to mass, the fundamental principle of dynamics (classically associated with the formula $\vec{F} = m\vec{a}$[30]) leads to an equation containing two terms proportional to this variable which then cancel out. Consider our skier again, but represented as a rectangular parallelepiped. Let's take another skier of the same shape, but much lighter (less dense). There again, do we imagine that these two skiers, who set off together, are going to arrive together at the bottom of the slope? The result's independence of mass is here merely the result of deliberately ignoring frictional forces, of magnitude $F_V$, due to the viscosity of air. These do not depend on mass, thus the motion does: dividing both sides by $m$ to get the acceleration, one term remains (the one associated with air friction) which includes this parameter in the denominator: $a = g(sin\theta - \mu_d\ cos\theta) - F_V / m$. The bigger the mass, the smaller this retarding term, $- F_V / m$, there is therefore some advantage in being a very dense champion!

Even though downhill skiing may not be our main interest in life, these thoughts allow us to preserve this general formulation: in elementary mechanics,[31] if all the forces exerted on the object whose motion is being analysed are proportional to its mass, then it is an irrelevant parameter; on the other hand, when some of these forces do not depend on it, the object's motion does depend on it. In *Explorers on the Moon* (the comic in the *Tintin* series by HERGÉ) why is the rocket, with its engines shut

30 Classical notation: total vector force $\vec{F}$ exerted on a "particle" of mass $m$; $\vec{a}$ is the acceleration of the particle.

31 The case considered here is the Galilean reference frame. In a system of two or more bodies, where the centre of mass is not necessarily identified with the most massive body, the situation gets complicated.

down, not drawn towards the asteroid Adonis, while Captain HADDOCK, on his ill-advised and inebriated space walk, is on a dangerous approach to it?[32] Certainly not because the poor captain is less massive than the rocket (with its engines silent and hence just as passive as him).

As for so-called "mass" spectrographs, these make use of forces due to an electric field which is not itself aware of the mass of the object being deflected, whence the influence of mass in the observed deflection. For the same reason, the path of a charged particle in a magnetic field depends very much on its mass.

## 2.6 The mirror

The field of a mirror is a classic exercise in the first years of university or in secondary school. All you need is DESCARTES' law in four strokes of a pencil (Fig. 2.3) and there you are, the problem is solved.

Solved maybe, but not finished, at least not in a very interesting way. If we decide to focus our attention on the part of the plane parallel to the mirror and visible to an eye in this plane,[33] this is the same thing as wondering what part of himself somebody can actually see in the mirror.

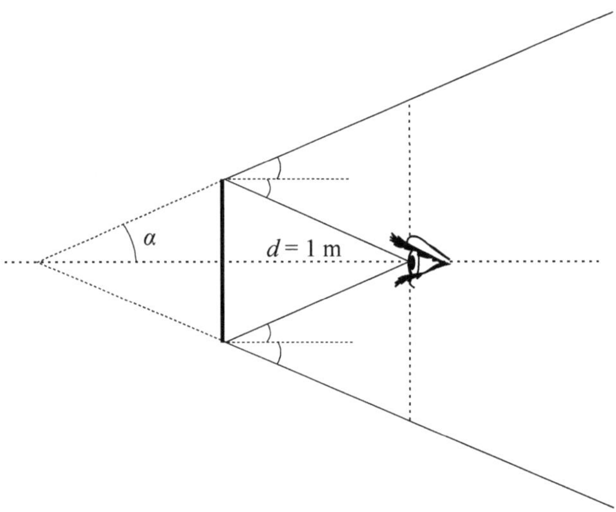


We can imagine that the answer expected by the teacher is as shown in this figure, based on the equality of the angles of incidence and reflection. For clarity, the scale is not the same in both dimensions.

***Figure 2.3*** *- A model answer for the question of a mirror's field. Let there be a plane circular mirror 10 cm in diameter. If you place your eye on the axis of the mirror and 1 m from it, draw a diagram to show the region of space you can see in this mirror.*

32 See HERGÉ, *On a marché sur la Lune* [Explorers on the moon], Dargaud, Paris 8.

33 Chapter 4, exercise 4.2. Concerning students' common ideas about mirrors, see GOLDBERG F. & MCDERMOTT L.C. (1986) Student Difficulties in Understanding Image Formation by a Plane Mirror, *The Physics Teacher*, **24** (8) 472-480; GALILI I., GOLDBERG F. & BENDALL S. (1991) Some reflections on plane mirrors and images, *The Physics Teacher*, **29** (7) 471-477.

We can do a miniature simulation of this situation by distributing small, 1 cm square mirrors in class,[34] together with small 4 cm cardboard squares with a hole in the centre and gridded in the same format as the mirrors. Looking through the central mirror, with the eye against the box and with the squares facing the mirror, we can count how many we see; this is the same as counting how many square centimetres of one's own face can be seen in the mirror. Typically, our first-year student replies: "four", and "me too" chime in the others. You don't hear immediately: "it doesn't depend on the distance!", that in other words the collection of individual results signifies an invariance. Little by little, one by one, you can see them experimenting, bringing their mirrors closer in and then further away. It is very striking to be able to literally read from people's movements, the sequence of intellectual operations taking place: measuring (as seen by the intensity of their concentration), then testing some possible invariance by changing the position of their arms.

Interesting, precisely because it is unexpected, this independence with respect to distance means that there is no point in going backwards to see your belt in the mirror if it has not already appeared in a first rough inspection.

As is often the case, the surprising invariance involves two factors which cancel each other out and for which only one was spontaneously taken into account.

As far as friction is concerned, one might think that the contact surface area is a determining factor, without necessarily realising that the crushing of asperities diminishes with increasing area. With the small mirror, we imagine that in stepping back more of one's own surface is fitted in a sort of cone of visibility subtended by the mirror, and whose apex would be at the location of the image of the eye in the mirror. However, this is to forget that the included angle of this cone ($\alpha$ in Fig. 2.3) decreases. THALES' theorem provides an exact compensation: the image of the eye is always twice as far from the apex of the cone as the mirror. Hence, the area of the plane of the eye which is still visible is always four times bigger than that of the mirror.

In terms of interest, is there anything in common between the presentation of the four lines in Figure 2.3 and what (with its rich additions) has just been described: collective experimentation,[35] invariance, discussion, a fresh look at a demonstration?

34 Like those on a disco (glitter) ball. Experiment suggested by W. KAMINSKI, private communication.

35 There is currently a consensus that the teacher should ask the students what they expect the outcome to be before performing any experiment. See for instance: WHITE R. & GUNSTONE R. (1992) *Probing Understanding*, London: Falmer Press. And, very recently: "The so-called clicker questions (the name comes from the electronic device that students use to record their answers) usually focus on common students misconceptions about the concepts (…)": "A transformed course typically begins not with a lecture but with a clicker question. Students gather in small groups to discuss it, and a fellow assigned to the course circulates through the classroom to guide the inquiry process. Once the students have punched in their answers, the faculty member might

To be sure, the familiar features of interest are also there: the "mini-demo", the discussion, and the link with everyday life. Why would we ignore them? Nevertheless, this does not invalidate the idea that highlighting an unexpected invariance is in itself a source of satisfaction, even though we cannot always analyse both the common initial expectation and the reason for the surprising result.[36]

## 2.7 The power and equivocalness of invariance

When we come across an unexpected invariance, how can we do otherwise but be intrigued? There, we have a vast array of situations for which, after dealing with one of them, we no longer have to invest our energy other than specifying precisely the field in question. Excessive generalisation leads to disappointment, but to do so insufficiently is to pay short shrift to the power of the solution. Although it occurs extremely frequently in physics, the function $y = Constant$, needs to be specified: constant with respect to which interesting variables and sensitive to variations in what others?[37] The two sides to this analysis will be looked at later from various different angles; repetition should not seem too annoying in the light of current practice. In particular, we shall see that what holds true for the value of a variable also applies to more qualitative statements. For instance, the following, very much in favour for the ideal gas law ($pV = nRT$, in the usual notation), to which we shall return in Chapter 3: "All gases which may be assumed to be perfect gases behave in the same manner". Obviously, it all rather depends on what is meant by "behave", and the risk of misunderstanding is ever present. Thermodynamics is not the only field in which such preoccupations are relevant.

A variable with a surprising lack of influence is noteworthy, but a pleasure to observe nonetheless.

---

offer a microlecture aimed at correcting their mistakes and filling in gaps in their knowledge. Once the concept is clear, the class moves on to the next clicker question.": MERVIS J. (2013) Transformation is Possible if University Really Cares, *Science*, **340** (6130) 292-296. The proposed handling of this situation is deliberately different: the aim here is the gradual emergence of the question of invariance, which might otherwise be short-circuited by the method usually suggested.

36 Note that it's a lot less easy in the case of the phase velocity of a wave.

37 See VIENNOT L. (2001) *Reasoning in Physics - The part of common sense*, Dordrecht: Kluwer Ac. Pub., Chapter 6.

# Chapter 3

# Analysis of functional dependence: a powerful tool

## 3.1 Numerical or functional?

As physics teachers, we are always going on about "formulae". Concentrating exclusively on this one aspect is frequently–and justifiably–decried, and we often hear people advocating the value of explanations "with no formulae".

Everyone knows that these "formulae" allow one quantity to be calculated from all the others in the formula. However, if that's as far as it goes, then there's not much more to be said for them. But, we can go beyond this simple calculation of the relationship between variables, assigning a value to one in a given situation defined by the others. It is more fruitful to consider also (or above all) the functional use of the relation: if one such variable increases, while the others save one remain the same, then will that last variable increase, decrease or stay the same?

Take NEWTON's second law: $\vec{F} = m\vec{a}$.

In a set of objects firmly attached to one another and experiencing equal acceleration, the one with the greatest mass is the one that experiences the greatest resultant force. However, the greater its acceleration, the greater the likelihood will be of a weak object breaking; and if extremely thin supports are involved[38] as in the case of the interior of a cryostat designed for extremely low temperatures on board a satellite rocket, then that presents a tricky problem at launch time.

Sometimes this kind of functional analysis can work the other way round. The formula can be found from an intuitive understanding of the dependencies. Let's imagine that we want to remember the expression for the radius of curvature $R$ of the path of a charged particle (of charge $q$ and mass $m$) in a magnetic field $\vec{B}$. We know that this involves a fraction, with two variables in the numerator and two others in

38 Shape memory materials can lead to relatively strong supports, retractable once weightlessness has been achieved; this can be interpreted as an absence of interaction between object and support both being immobile within the reference frame of the station (this system was used on the PLANCK satellite launched in May 2009, whose mission was the directional analysis of the cosmic microwave background).

the denominator. Is it $qB$ over $mv$ or $mv$ over $qB$? A complete blank! A swift analysis of what contributes to the particle's inertia ($m$ and $v$) and of what is likely to be involved in the forces determining the curvature ($q$ and $B$), and remembering that a large radius of curvature corresponds to something that doesn't turn much, then all doubt disappears: $R = \frac{mv}{qB}$.

Using such elementary examples, it is this type of reasoning which comes into play when taking a functional view, and its value is well recognised by the great majority of physics teachers. It is all the more important to come back to this as, what is at stake is the possibility of improving their chances of becoming strongly involved in current teaching practice, and also of highlighting one or two potential traps.

## 3.2 Before physical values: the relationship

There is an exercise which is a classic among classics:[39] namely, the calculation of the radius of curvature ($R$) of the path of a charged particle (of charge $q$) in a magnetic field $\vec{B}$. The expression for the so-called Lorentz force exerted on the particle is $\vec{F} = q(\vec{v} \times \vec{B})$. Before considering the values of the variables we observe that the force is always perpendicular to the velocity vector $\vec{v}$ (and hence the path) as a consequence of the vector product.

In this model a magnetic field cannot alter the linear velocity of the particle. Only the direction of motion can be affected by the field. It is remarkable that so many trainee teachers, not to mention students, find the following question very disconcerting (given in an answer sheet in the case where $\vec{B}$ is uniform): if the field $\vec{B}$ is no longer uniform, is the linear velocity still constant?[40] As we have just seen, the expression for the force justifies an affirmative response, meaning that no magnetic field can increase or decrease the velocity of a charged particle. All the same, the wording in fact provides useful information, even though the student might get the maximum mark for the standard exercise without ever having explicitly stated this powerful conclusion. The invariance of a result or the generality of a conclusion may seem trivial to us, but is not necessarily so for the student.

As far as the form of relationships is concerned, it's a good thing that some are appropriately highlighted in current teaching practice, for instance the linear relationships between vector variables, primarily in elementary mechanics. Writing down NEWTON's second law with a constant force term comes down to preparing

39 See Chapter 4, exercise 4.3.

40 This question, which requires only "well known" things, is mentioned here as a result of personal experience over fifteen years of instructing trainee teachers (roughly 350 individuals): except for a few very rare cases, the trainees had never heard the question before, and were often perplexed as to what answer they should give. One might even ask if the relation remains valid in the case of a non uniform field.

a solution in which there is separation of each spatial dimension. There we see the power of theory, reduced as it is. Hence we know, with no calculation involved, that a force acting in a single direction ($\overrightarrow{0y}$ for example) will have no effect on the path of a moving object in a direction ($\overrightarrow{0x}$).[41]

Whether these are standard examples or not, this type of conclusion can never be emphasised enough.

## 3.3 Keeping an eye on a causal reading of relations

Contrary to received wisdom, the inclusion of variables in a relationship does not necessarily correspond simply to a causal analysis of the situation.

Hence the situation Marie CURIE suggested to her young students[42] of a small ball immersed in a bowl of water. Disturbingly, her student Isabelle CHAVANNES' notes state the following:

"*What was exerting pressure on the ball when it was in the water? The water, of course,* ***but also the air, which was itself pressing on the water. This air pressure was transmitted through the water.*** *When the ball was on the surface of the water, only atmospheric pressure was pressing on it; when I pushed it under the water, it had to support* ***both the atmospheric pressure and*** *the pressure of the water.*"

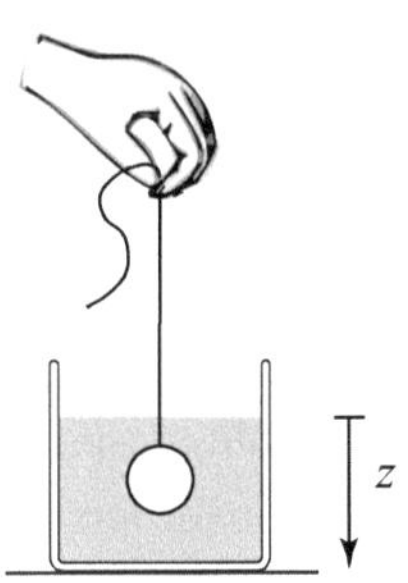


***Figure 3.1*** *- Small ball immersed in a bowl of water.*

*This simple situation can be read in two ways.*

- *A causal interpretation of hydrostatic pressure. Marie CURIE's explanation as reported by a student. What was exerting pressure on the ball when it was in the water? The water, of course, but also the air pressing on the water. This pressure is transmitted through the water;*
- *A Newtonian reading of the situation: what was exerting pressure on the ball when it was in the water? The water.*

*In the expression* $p(z) = p_{atm} + \rho g z$, *there are two terms in the expression for* $p(z)$, *but this variable locally characterises the water and determines its interaction with the ball.*

---

41 An exercise suggested during a study of French *Terminale* (Year 13) students features this property for a mass spectrometer: does the transit time for a particle of given charge and initial velocity parallel to the plates of a plane capacitor depend on the fact that it is charged or not? Besides the significant number of errors, the students' answers provided interesting food for thought. (RIGAUT M. & VIENNOT L. (2002) Réduire le théorème du centre d'inertie : jusqu'où ? *Bulletin de l'Union des Physiciens,* **841**, 419-426).

42 Collected *Lessons of Marie CURIE*, Isabelle CHAVANNES (1907) *Physique élémentaire pour les enfants de nos amis*. Work coordinated by B. LECLERCQ (2003), EDP Sciences, Paris, p. 33

If the parts in bold fonts of this reasoning are removed, the text is perfectly consistent. On the surface of the ball (submerged to a depth $z$) the water exerts contact forces due to the pressure determined by the expression $p = p_0 + \rho g z$ (using the usual notation for variables, axis $\overrightarrow{0z}$ directed downwards, and origin at the surface of the water). Even though this expression includes two terms involving atmospheric pressure $p_0$ and depth of immersion $z$ respectively, it is the water, and only the water, which exerts contact forces on the outside of the ball. A causal reading must not allow the strict meaning of the expression to be forgotten.

At a higher level of competence,[43] there is one case where the relationship disturbingly hides the factor that determines the value of a variable. This is the expression for the electric field $\vec{E}$ in the neighbourhood of a conductor in electrostatic equilibrium: $\vec{E} = \frac{\sigma}{\varepsilon_0}\vec{n}$, where $\vec{n}$ is a unit vector normal to the conductor directed outward to the point under consideration, σ is the local surface charge density and $\varepsilon_0$ is the permittivity of free space (Fig. 3.2).

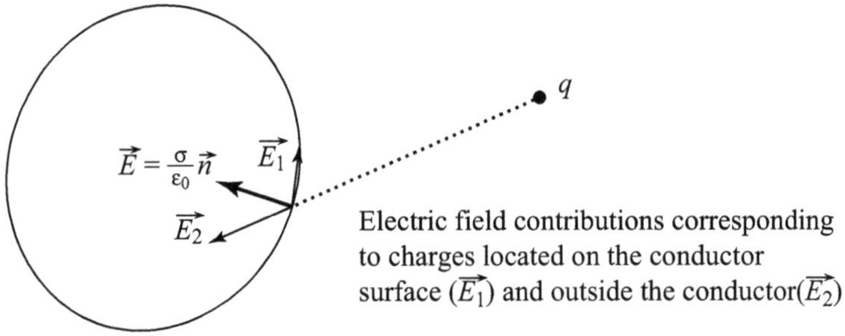


***Figure 3.2*** *- The electric field in the neighbourhood of a conductor is normal to it. So far as the charge is concerned, the expression mentions only the surface density σ in the neighbourhood of the point being considered, but in fact this field results from the contribution of all charges in the universe (here a single external positive charge is shown, while σ is negative).*

If we ask students what are the sources of this field $\vec{E}$, the overwhelming response is that it's down to the individual charges on the conductor (locally, or over the whole conductor).[44] However, the principle of superposition means that the field at some arbitrary point of some arbitrary configuration is the sum of the contributions from *all* the charges in the universe. Should the only source admissible by the students appear in the formula?[45] Within the variable σ is the cumulative contribution of all the charges present, both outside the conductor and at its surface. The power and

43 Typically in France this would be the second year at university or a preparatory class for entry into a *grande école*, second year.

44 See Viennot L. & Rainson S. (1999) Design and evaluation of a research-based teaching sequence: The superposition of electrics fields. *International Journal of Science Education*, Special issue: Conceptual Development in Science Education (continued), **21** (1) 1-16.

45 In the work of S. Rainson (previous footnote), the "cause in the formula" syndrome is mentioned in this connection.

unintuitive nature of the theorems of electrostatics are nicely illustrated here, and it is somewhat comforting that such a reduced expression takes account of such potentially varied situations.

The relationships between physical quantities on an excessively direct causal reading must be considered with some care. This observation might usefully be extended to statistical expressions unduly interpreted as if "correlation" meant "causal relationship". But that's another story.

## 3.4 Some factors not apparent (but by no means always trivial) in a relation between quantities

The risk to which we now draw attention was clear in what was said above. In certain cases the variables involved in a phenomenon may be ignored simply because an important formula does not contain them; as, for example, with the molecular mass of a gas. For the sake of simplicity, let us consider this gas as "perfect"; and in fact, such a state is very often not far from the truth. We can use the famous formula $pV = nRT$, with the usual notation.[46] However, the perfect gas relation might induce us to suppose that the molecular mass is of no importance.[47] However, at any given temperature the average speed of the molecules depends on the molecular mass of the gas:[48] speaking in a helium atmosphere, which is a gas of low molecular mass, creates a "Donald Duck" voice as the average molecular speed is greater than that of air under the same conditions. Likewise, helium diffuses faster than any other so-called "heavier" gas at the same temperature. So much for the kinetic aspect. As for the behaviour of a gas in a gravitational field, we can expect the molecular mass to be involved. Helium is used for balloons because it is a "light" gas. The properties of the atmosphere, even its very existence, are very much related to this often ignored parameter, the molecular mass of the gas. All this should help put into perspective

---

46 $p$ : pression, $V$ : volume, $n$ number of moles, *i.e* the number of molecules $N$ divided by AVOGADRO's number ($A = 6.023.10^{23}$), $R$ is the ideal gas constant and $T$ is absolute temperature.

47 On this topic see CHAUVET F. (2004) Une simulation pour explorer un modèle cinétique de gaz en seconde, *Bulletin de l'Union des Physiciens* **98** (866) 1091-1105, as well as his online training documents (http://www.epi.asso.fr/revue/articles/a0306d/Gaz_a.htm). See also: KAUTZ C.H., HERON P.R.L., LOVERUDE M.E. & MCDERMOTT L.C. (2005) Student understanding of the ideal gas law, Part I: A macroscopic perspective, *American Journal of Physics,* **73** (11) 1055-1063; KAUTZ C.H., HERON P.R.L., SHAFFER P.S. & MCDERMOTT L.C. (2005) Student understanding of the ideal gas law, Part II: A microscopic perspective, *American Journal of Physics*, **73** (11) 1064-1071.

48 The classical relations show that, at a given temperature $T$, the mean square speed $\overline{v^2}$ is inversely proportional to the molecular mass $m$: $\overline{e_c} = \frac{1}{2} m\, \overline{v^2} = \frac{3}{2} kT$, where $\overline{e_c}$ is the particular mean kinetic energy and $k$ is BOLTZMANN's constant ($k = R\frac{n}{N} = \frac{R}{A}$). For notation, see note 46.

the dictum so widespread in schools and loosely expressed as "all gases behave identically at low pressures".[49]

While it is important to emphasise the importance of these laws and their impressive generality, a discerning look at what they really say can avoid embarrassing paradoxes.

## 3.5 Functional dependencies and graphs: an example in geometrical optics

Saying that functions can usefully be represented by graphs may seem obvious. As with so many principles alluded to in this book, that's nothing new, so we need an example to justify coming back to this. One example illustrates how the distance between theory and practice could be reduced and provides if not a reason, at least an occasion, to reconsider the dividing line between what we do by habit and what we affirm will happen. To keep things simple, our example comes from elementary geometrical optics.

Under so-called "Gaussian" conditions (which should be carefully reviewed here), there is an expression which relates the position of a point object ($A$), a source of light, and that of its image ($A'$) formed by a thin lens or spherical mirror. On an axis (perpendicular to the lens or mirror) traditionally oriented in the assumed direction of the incident light (from left to right on the diagram), and from centre $O$ of the lens or maximum $S$ of the mirror, the abscissae of these positions are denoted respectively by $p$ (or $\overline{OA}$ or $\overline{SA}$), and by $p'$ (or $\overline{OA'}$, or $\overline{SA'}$). The focal points denoted by $F$ and $F'$ are associated with the focal distances $f = \overline{OF}$ and $f' = \overline{OF'}$ or their equivalent for mirrors. Familiarity with this subject presupposes that the formulae linking these algebraic variables are well known. Whether for a convergent or divergent lens, there is a single formula: $\frac{1}{p'} - \frac{1}{p} = \frac{1}{f'}$ but a veritable mine of applications.

Depending on the signs of the variables involved, the various cases in the figure come in a multitude of forms: real image ($p' > 0$), virtual image ($p' < 0$), for a convergent ($f' > 0$) or divergent ($f' < 0$) lens and depending on whether the object is at such and such a position with respect to the focal point. There is a whole host of variants. The collection of corresponding diagrams is a poor incentive for memorising them, but they can be reproduced by the use of simple rules governing the rays. For mirrors,

49 The influence of the molecular mass of a gas and even this parameter itself are not mentioned at all in the syllabus for the *classe de Seconde* (Year 11) in use in France for the decade 2000-2010 (*MEN-Bulletin Officiel,* Spécial édition N°12, Août 1999). The "slogan" mentioned here is similar to phrases found in school books such as *Physique Seconde*, Hachette Education, 2000, DURANDEAU *et al.*, p. 124: "At low pressures all gases exhibit identical behaviour, this being the characteristic of a so-called ideal or perfect gas". The vague school-level idea of 'thermal agitation' as something which might be related to temperature muddies the waters: it's actually the average kinetic energy per particle which is directly related to temperature (see note above).

whether concave ($f' < 0$) or convex ($f' > 0$), just one sign change in the formula will suffice: $\frac{1}{p'} + \frac{1}{p} = \frac{1}{f'}$ and the previous statements are transposed.

The basic procedure in standard exercises is to apply the appropriate formula to calculate one variable from the other two. When reciprocals are involved, it can be tiresome, since it is easy to make a numerical mistake. Not much can be done about poor calculating skills, apart from attempting a time consuming geometrical construction, and a purely numerical solution is rather dull. There are quite a few functional statements going around in current practice, e.g.: "if the object is displaced from left to right, so is the image formed by a convergent lens". Fine, and what about a divergent lens? Apart from the specialists in the field, no-one knows.

We suggest a more obviously functional approach via a transformation of the previous formulae. Taking the origins of the abscissae at the focal points ($\overline{FA}$ or $\overline{F'A'}$) gives us what is known as NEWTON's formulae.[50]

The object-image correspondence in the lenses is then summarised in the relation:

$$\overline{FA} \, . \, \overline{F'A'} = -\, f'^2$$

and for mirrors (focal points $F$ and $F'$ are then identical):

$$\overline{FA} \, . \, \overline{F'A'} = f'^2$$

There are no more reciprocals, just two algebraic variables (let's call them $d$ and $d'$) whose product is constant. A graph of the relationship between them has the form of a hyperbola centred at the origin (for mirrors, $d.d' > 0$) or of a symmetric curve with respect to any one of the axes (for lenses, $d.d' < 0$). If we wish to reuse the variables $p$ and $p'$, we have only to carry out an inverse change of variable: $p = d - f'$, $p' = d' + f'$.

The graphical interpretation of such changes of variable is a displacement of the appropriate hyperbola parallel to each axis by an amount $-f'$ or $f'$ depending on the particular case. The result is given in Figures 3.3 and 3.4, which summarise all the positional correspondences for thin lenses and spherical mirrors using the GAUSS approximation.

50 The equivalence for lenses is demonstrated by the following equivalent expressions: $\frac{1}{p'} - \frac{1}{p} = \frac{1}{f'}$
*i.e.*: $(p - p') \,.\, f' = pp'$ *i.e.*: $-f'^2 + (p' - p)f' + pp' = -f'^2$
*i.e.*: $(p' - f')\,(p + f') = -f'^2$ or $\overline{FA} \,.\, \overline{F'A'} = -f'^2$

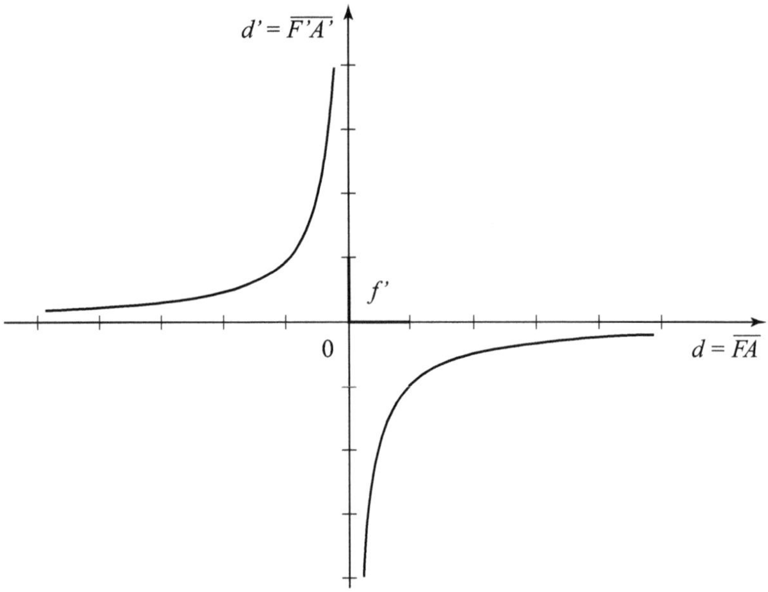


***Figure 3.3*** - *Relationship between the abscissae of the object and the image for a convergent thin lens (f'>0), with origin at the focal points.*

Over the course of the lecture these figures can be reconstructed using two transparencies, one with a centralised hyperbola, and the other simply with two mutually perpendicular axes graduated in values of the focal length $f'$. As far as positions are concerned, shifting the transparencies around miraculously produces the whole of elementary geometrical optics, just with two very simple transparent diagrams.

The effect is guaranteed, for who would read a graph and associate a change of origin with a change of variable as a “simple displacement”?

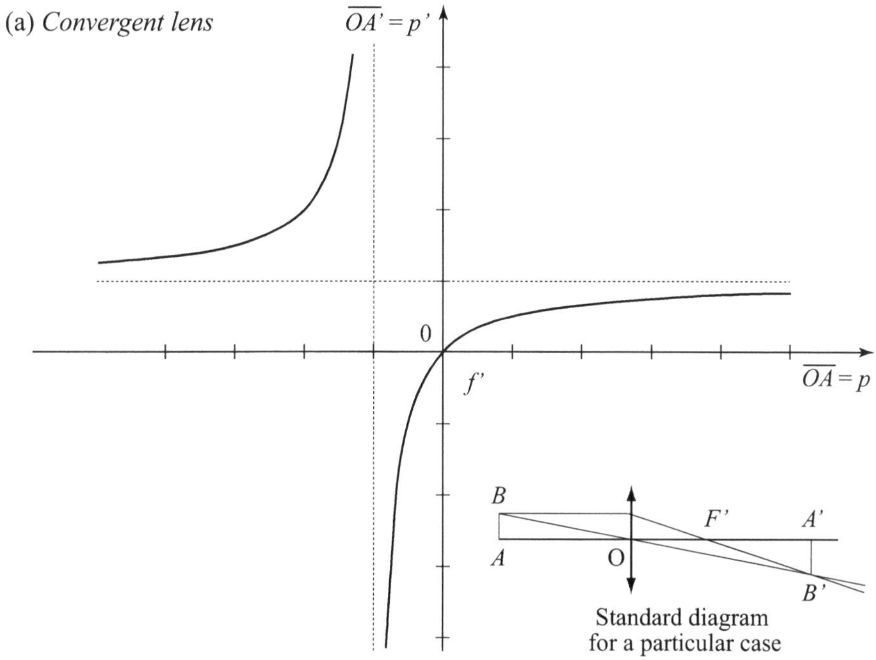


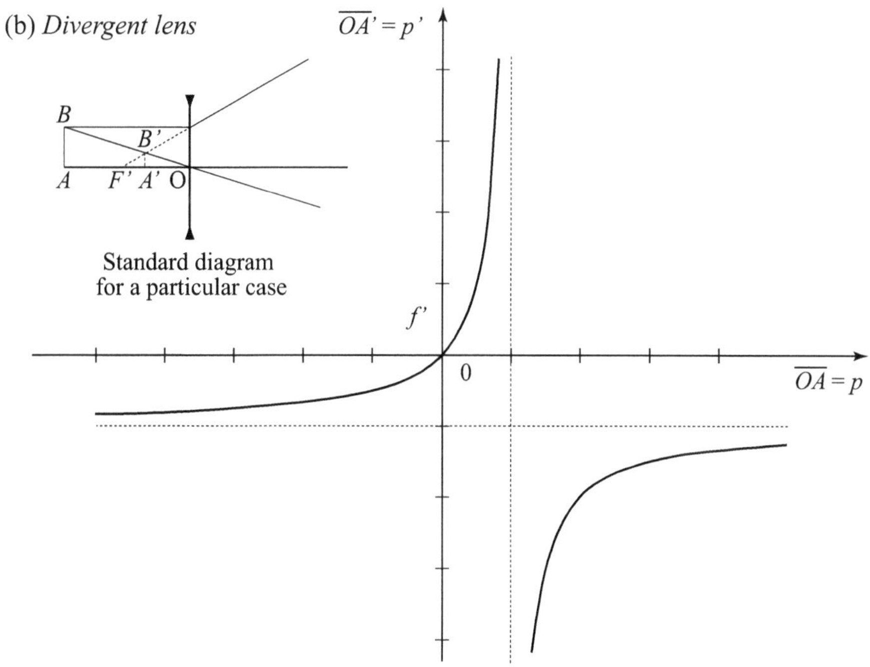


***Figure 3.4** - Conjugation relations for thin lenses, for* (a) *a convergent lens,* (b) *a divergent lens. Standard diagrams are shown beside the graphs p'(p)* ($p = \overline{OA}$, $p' = \overline{OA'}$).

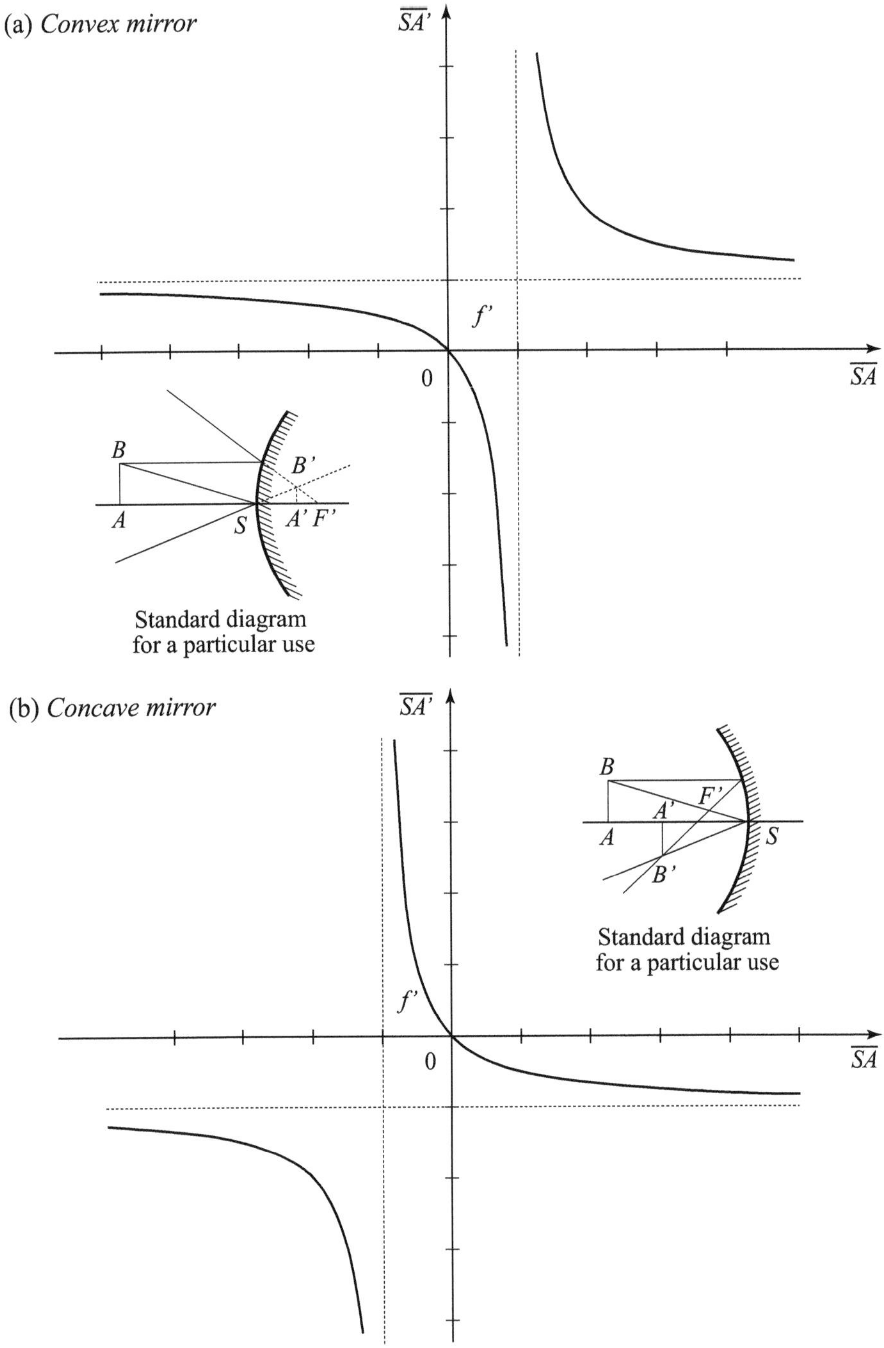


***Figure 3.5** - Conjugation relations for spherical mirrors in case* (a) *for a convex mirror, and* (b) *for a concave mirror. Standard diagrams are shown beside the graphs* $\overline{SA'}$ $(\overline{SA})$.

We can get a great deal of satisfaction just from having such a concise, amenable and easily memorised summary of the field concerned, but what merits its place in this chapter is its functional utility. Returning to this narrative of object and image which (for the sake of brevity) move in the same direction with a thin lens. In functional terms this means that the function $p'(p)$ is increasing, as is shown in Figure 3.4. However, it now clear that this is also true for a divergent lens. The mirrors are associated with a decreasing function $p'(p)$ (Fig. 3.5), and they therefore displace the image in a direction opposite to the displacement of the object on the axis. The plane mirror is no exception, as illustrated by the limiting form of the formula ($p' = -p$) for an infinite focal length, and the associated graph (the second bisector).

The most trivial of optical exercises can be transformed so as to capitalise on the merits of such graphs. In the following pages[51] the reader can compare a standard version with its reformulation, matched up with their respective answer sheets. It seems at the very least that the corresponding intellectual processes are different. It could easily be concluded that for this situation the ideal would be to solve both versions of the exercise, and then compare and summarise the two approaches. For the moment, we can at least observe that the range of possible actions is more open-ended than current practice would suggest, noting the cumulative effort in abstraction required by the functional approach illustrated here, and a general familiarity with graphs.

## 3.6 Some neglected treasures and hazards highlighted

As far as the functional approach is concerned, the one or two items reviewed in this chapter make no claim to do anything more than highlight the usual injunctions more effectively. Indeed, in relation to the situations discussed here, these injunctions are rarely honoured. For example, we could even say that with respect to geometrical optics, it is exceedingly rare despite the patently obvious benefits, we are left wondering why. To say that the functional approach is swamped by conventional teaching rituals is no explanation, and is just a reformulation of this rarity. It is true that geometrical conjugation diagrams have a prominent place in the exercises routinely set in optics with their associated panoply of tedious elementary calculations. However, it is also worth considering an approach which brings with it a number of additional benefits.

As for the necessary vigilance with regard to the potential disappearance of variables resulting from such and such a calculation, there is a problem of a different nature:

51 Chapter 4, exercise 4.5. On assessing the practical aspects of graphs in the baccalauréat (1999 and 2000) see M. Rigaut's thesis (2005, in French): *L'épreuve écrite de physique au baccalauréat: analyse du point de vue du contrat didactique*, a study centred on the years 1999 and 2000, Université Paris 7 (www.matthieurigaut.net/public/docs/these_didactique_matthieu_rigaut.pdf).

that of linguistic shortcuts in the matter of functional dependence. To "depend" or "not to depend" on such a variable is often indicative of an inappropriate formulation, since it may be incomplete. "The molecular mean square speed of a gas depends only on its temperature" is a statement deemed true[52] for a gas of given composition, but false in the case where it leads to comparisons between gases of different compositions. Rather than just hoping that no ambiguous statement will be made, it is better to concentrate on making the meaning clear, thereby avoiding the risks to which everybody is exposed.

The examples in this chapter are essentially intended to take *explicit* account of the inner workings of the physicist's analysis, here in its functional guise.

52 In the classical approximation, but without requiring the gas to be perfect: Diu B. *et al.* (1989) *Mécanique Statistique*, Hermann, Paris, 350-352.

# Chapter 4

## PUTTING THINGS INTO PRACTICE

### 4.1 Introduction

The texts which follow illustrate the variations which can accompany a standard exercise, drawing inspiration from some of the previous thoughts. There is no attempt to provide the reader with some sort of ideal formulation, and it will doubtless be possible to find fault with some of the suggestions here. The overriding aim, however, is to open up the range of possibilities within a typical school or university framework. The suggestions here have all been put into practice by the author, and by teaching teams in first year university physics courses.

Three of the texts in this chapter are direct illustrations of the critical use of answer sheets.

The rationale behind this kind of proposal is as follows: knowing how to understand the work of others (in relation to science) is a generally useful skill, not just for the future citizen, the pupil copying his classmate, the student studying a course, but also for the future researcher. The critical examination of a text is not something which can be improvised, but has to be properly taught.

The approach adopted here is that of analysing the ordinary answer sheet of a standard exercise: there are examples taken from university first year tutorial sheets or textbooks, and from the *Terminale* (Year 13). As for the answer sheet taken as the basis for the proposed work, the "text" used can be reduced to a calculation framework or diagram, or can include several verbal links. The aim is not to assess the merits of such "texts" but to extend their use by stimulating thought processes.

When working within this framework, it may come as rather a surprise to find that students should have an answer sheet available immediately the exercise is set. Of course, this is by no means an obligatory scenario, but put forward as a powerful illustration of the difference between a process that involves critical thinking and a solution-seeking process. It is quite clear that, ideally, these two ways of working should be associated, but focussing on just one of them has the advantage that, freed from the agony of calculation, the student need only reflect on what he has in front of him. It takes the same time to work with this exercise but in another way.

Experimentation on this unusual exercise was carried out and positively assessed in the *Terminale* (Year 13), and also in tutorials over two weeks in an entire section at Paris 7 university (about 120 students in first year). A series of other explanatory suggestions for the text can be found in the documents relating to these two experiments,[53] in which a series of question types is suggested and illustrated. Without being dogmatic about it or making a restrictive list, it is at least some source of inspiration. The topics or direct suggestions for questions are as follows:

## Topics or suggestions for questions

- The meaning of symbols or verbal expressions used in the text being analysed.
- The influence of "coding" in designating symbols and in writing down algebraic relationships in the text: this section relates to what is often described as the "sign convention". One might wonder how the way a calculation is written alters when the orientation of an axis is changed (hence "*z* axis").
- The assumptions made in the text: here we examine the very foundations of the argument. This section deals with questions which frequently prove difficult, for example:
  - Where does such an assumption come in?
  - On what assumption is such and such an assertion based?
  - What must be altered if such an assumption is changed?
  - Analyse the orders of magnitude which justify such an assumption.
- Calculation, thus:
  - Is such and such an equality in fact proved in the text from some other one, or is it simply "parachuted in", e.g. "... it is well known that..."?
  - Where are the simple intermediate steps of the calculation?
- The result: what is involved here is the significance of the conclusion, hence:
  - Give a full written statement of the result
  - Check the result: consistency and limiting cases

53 Groupe A. Cros (1983) Les exercices en classe terminale, *Bulletin de l'Union des Physiciens*, **659**, 385-416; Viennot L. (1987) *Corrigés : mode d'emploi* [answer sheets and how to use them], Université Paris Diderot (Paris 7), LDSP (now called the *Laboratoire de Didactique André Revuz*).

- How does a given variable vary in cases where another increases, the others remaining unchanged?
- Obtain an order of magnitude for such a variable under specified circumstances.

• Extension: slightly extend the calculation given in the text to draw out further information.

We can immediately recognise here, this pursuit for meaning which the previous chapters were aiming for, and especially this preoccupation with developing a functional analysis over and above a mere exercise in numerical relationships, limited to "finding the formula" for calculating a result. The embarrassing and fruitful questions about the assumptions unite with the topics developed above, the key thing being the discovery of invariants and the possibilities for generalisation.

Consistent with the simplicity which marks this book, the examples given here concern so-called elementary physics. Rather than just producing a list, we aim to illustrate a process. These texts deal with:

- the field of a mirror;
- deflection of a particle in a magnetic field;
- a block sliding on an inclined plane;
- a slide projector;
- flotation between two immiscible liquids.

These embrace the issues already used in developing the reflections above, and will not be detailed again here. Neither are the "expected answers" provided any more comprehensive, since the reader will be well aware of the classic solutions to such exercises.

## 4.2 The field of a mirror

The field of a mirror has already been covered in Chapter 2. Initially we may give the students the statement of the exercise and its standard solution. We then ask questions to stimulate interaction or enhance the scenario as described in 2.6. Phrases in italics are intended for the teacher.

### Statement of the question and answer sheet in a current version

Let there be a plane circular mirror of diameter 10 cm. If you place your eye at 1 m from the mirror on its axis, draw a diagram to show the region of space you can see in this mirror.

It is probable that the answer expected by the teacher would be the kind of drawing shown in Figure 4.1, based on the equality of the angles of incidence and reflection. For convenience, the scale is not the same in the two dimensions.

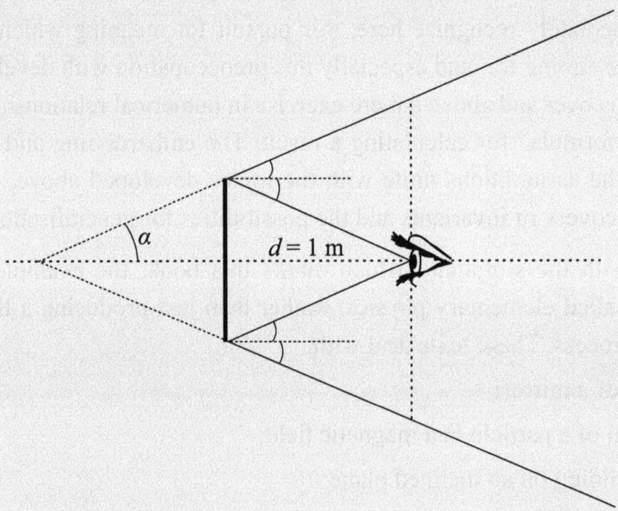


***Figure 4.1** - A standard answer to the mirror field question.*

### Suggestion for deeper investigation

*In order to get a fuller understanding of what is involved, the student could be asked the following questions (as described in the text, Chapter 2) after providing him/her with both the text for the exercise and the above suggested solution.*

- Which variable(s) can be used to characterise the "region of space" we are interested in here?

*The following variables can be used: the apex angle α of the cone restricting the visible space; the corresponding solid angle; the diameter of the part visible from the observer. Their relative merits and interrelationships can be discussed.*

How does the region of space in question vary when you move your eye further back from the mirror? Can the observer see a bigger portion of himself?

*Assuming we have a circle as if the observer were flat,* THALES*'theorem gives a diameter double that of the mirror and hence four times the area. For example, see Figure 4.2 which, compared with Figure 4.1, leaves the length of the vertical dotted line at the eye unchanged.*

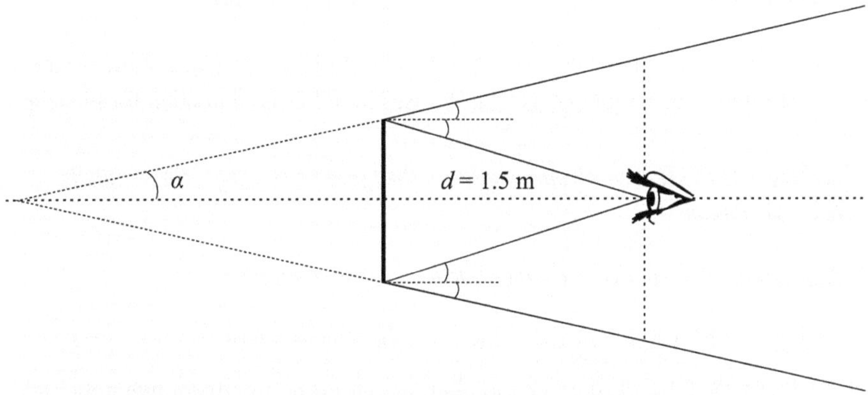


***Figure 4.2** - Altering the distance between eye and mirror does not change the part of the vertical plane of the eye visible in the mirror (see Figure 4.1): its area remains four times that of the mirror.*

– It's essentially a question of seeing. In the first diagram, it is not clear how the observer can see an object. Fill it in to include the path of a beam of light so that the observer can see an insect bite on his face. Is the beam convergent or divergent?

*Despite the usual "convergent beam" response, reinforced by a superficial examination of Figure 4.1, the beam is actually divergent. The diagram in Figure 4.3 highlights the importance of the width of the pupil: you can't see much "with a single ray" (besides, this expression is, of course, meaningless!) For an even more advanced analysis, you can associate with this the fact that the limits on the field of visibility do not correspond to a discontinuity.*

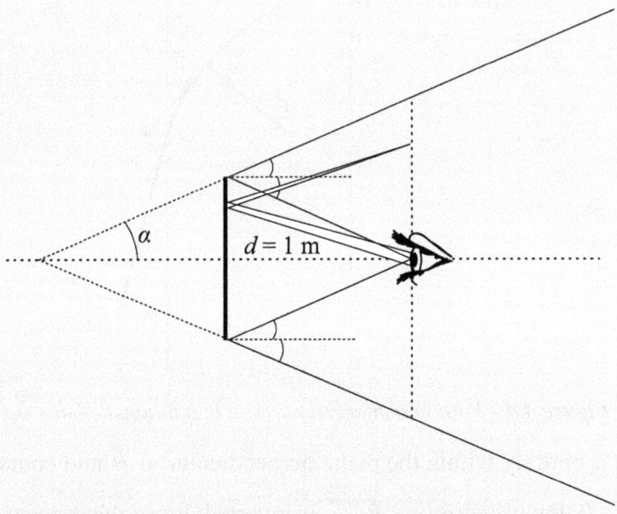


***Figure 4.3** - In the vertical plane of the eye, the light scattered by an insect bite penetrates the pupil in a divergent beam*

## 4.3 Deflection of a charged particle by a magnetic field

This topic has already been covered in 3.2. What follows is a suggestion for the text to be submitted to the students, as it is. Phrases in italics are intended for the teacher.

***Carefully read this text which appeared published several years ago in a Terminale (Year 13) manual:***

***Investigation of the path of a charged particle in a magnetic field***

Field $\vec{B}$ is uniform; the initial velocity $\vec{v}_0$ is orthogonal to $\vec{B}$.

Consider a particle of mass $m$ and charge $q$ entering a uniform magnetic field $\vec{B}$. The initial velocity $\vec{v}_0$ is orthogonal to the magnetic field lines. Neglect the mass of the particle; it can thus be assumed that the only force acting on it is electromagnetic in origin:

$$\vec{F} = q(\vec{v} \times \vec{B})$$

The acceleration $\vec{a}$ is related to the applied force by the fundamental relation of dynamics:

$$m\vec{a} = q(\vec{v} \times \vec{B})$$

At any instant, vector $\vec{a}$ is orthogonal to $\vec{B}$.

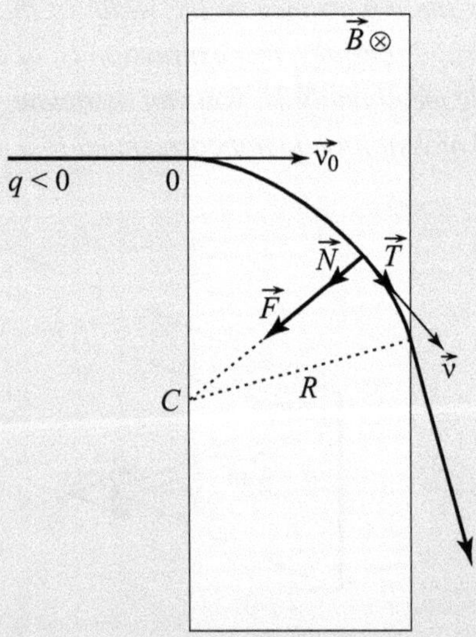


***Figure 4.4** - Path of a charged particle in a magnetic field $\vec{B}$.*

The path lies entirely within the plane perpendicular to $\vec{B}$ and containing $\vec{v}_0$.

Suppose $q < 0$: the trihedron $\vec{v}$, $\vec{B}$, $\vec{F}$ is inverted. From this we can deduce the direction of the deflection.

In the FRENET reference frame we have:

$$\vec{F} = m(\overrightarrow{a_N} + \overrightarrow{a_T})$$

$$= m\frac{v^2}{\rho}\vec{N} + m\frac{dv}{dt}\vec{T} = q(\vec{v}\times\vec{B})$$

and hence:

$$\vec{F}.\vec{v} = 0 \longrightarrow \overrightarrow{a_T} = \vec{0}$$

$$\frac{dv}{dt} = 0 \longrightarrow v = v_0 = Cte$$

$$q(\vec{v}\times\vec{B}) = \frac{mv^2}{\rho}\vec{N}$$

where ρ is the radius of curvature of the path.

The relationship between magnitudes and absolute values is written:

$$|q|\,v\,B\sin\alpha = \frac{mv^2}{\rho}$$

Put $v = v_0$ and $\sin\alpha = 1 \quad (\alpha = 90°)$

hence $|q|\,v_0\,B = \frac{mv_0^2}{\rho} \longrightarrow \rho = \frac{mv_0}{|q|B}$

The radius of curvature is constant, and the path is a circle of radius

$$R = \frac{mv_0}{|q|B}$$

## Then answer the following questions

*(By way of example, there is a list of possible questions with overlap among the answers: it is up to the reader to choose or reformulate them.)*

- What do the symbols $\vec{a}_N$, $\vec{a}_T$ and $\alpha$ represent?
- "Neglect the mass of the particle": with respect to what? Imagine the zone of a mass spectrometer where, before striking the detector, the particle is subjected to no force other than its own weight. What can then be neglected in order to treat the path as a part of a straight line?
- How does the path alter if the sign of the charge changes? Analyse the way in which the calculation is written down algebraically: is it necessary to use the absolute value notation, and can the variable ρ be negative and still retain meaning in the calculation?
- How does the radius of curvature vary with each of the (other) parameters? From this discussion create a mnemonic to avoid inverting the fraction which gives this radius of curvature.
- State the property of the force which allows you to assert that the motion is uniform.
- Is the motion still uniform
  - if the magnetic field $\vec{B}$ is no longer uniform?

- if the magnetic field is uniform and the initial velocity is not perpendicular to $\vec{B}$?

– What assumption(s) is/are made in the text which may lead to the assertion that the particle is confined to a plane?

– What is the motion of the particle if $\vec{v}_0$ is parallel to the uniform field $\vec{B}$? Is the path stable?

– The fact that the field $\vec{B}$ is uniform is mentioned several times in this demonstration. Recapitulate where and how.

– Give orders of magnitude for the values of $v$, $B$ and $R$ for the LHC (Large Hadron Collider at CERN, Geneva)?

## 4.4 Sliding on an inclined plane

This topic has already been covered in Section 2.4. What follows is a suggestion for the text to be submitted to the students, as it is. Phrases in *italics* are intended for the teacher.

***First read (the first part of) this exercise and its answer sheet: the skier, in a customary version***

***Ski jump***

A skier of mass $m$ descends a piste consisting of:

– a rectilinear section AB making an angle $\theta$ with the horizontal and of length $AB = L_1$.

– a horizontal rectilinear section BC of length $L_2$.

Denote the magnitude of gravitational acceleration as $g$. Assume that there is friction between the skis and the snow: let $\mu_s$ and $\mu_d$ be the static and dynamic coefficients of friction respectively.

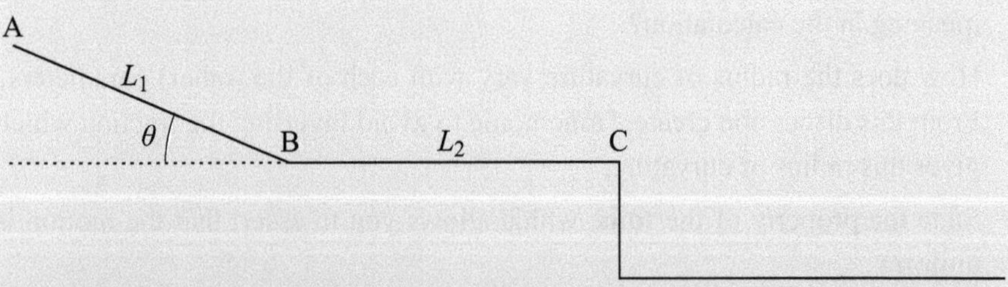


***Figure 4.5** - The elements of a ski slope involved in the ski jump problema)*

a) The skier is at rest on section AB. Draw a diagram showing the forces exerted on the skier. Give the (absolute) value of the frictional force as a function of $\theta$. So that the skier can remain stationary on the slope (without using his ski-poles), $\theta$ must be less than a maximum value $\theta_0$. Give an expression for $\theta_0$.

b) Now assume that the angle $\theta$ is sufficient for the skis to slide on the snow. The skier starts from A with speed zero at $t = 0$. Determine the skier's acceleration.

c) Choosing an axis $\overrightarrow{0x}$ coincident with AB and with its origin at A, determine the time dependent equation of motion.

d) Let the time taken by the skier to reach point B be $t_1$. Determine the value of $\mu_d$ as a function of $g$, $\theta$, $L_1$ and $t_1$. Give the value of the speed $v_1$ of the skier at point B as a function of $L_1$ and $t_1$.

The teacher's notes include the diagram below and the solution lines which follow. Read this carefully (there are no errors in the calculations).

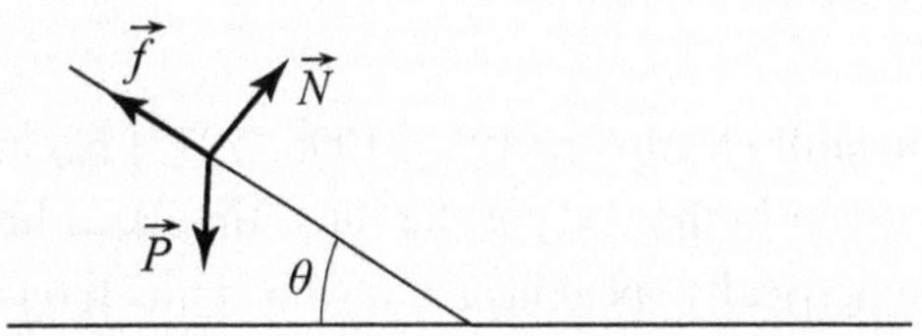


***Figure 4.6** - Diagram for the simplified skier model*

a)
$$f = \|\vec{f}\| = mg \sin\theta$$
$$f < \mu_s\, mg \cos\theta \longrightarrow \theta < \theta_0 \text{ with } \tan\theta_0 = \mu_s$$

b)
$$f = \mu_d\, mg \cos\theta$$
$$ma = mg \sin\theta - \mu_d mg \cos\theta$$
$$a = g\,[\sin\theta - \mu_d \cos\theta]$$

c)
$$x = \frac{1}{2} g\,[\sin\theta - \mu_d \cos\theta]\, t^2$$

d)
$$L_1 = \frac{1}{2} g\,[\sin\theta - \mu_d \cos\theta]\, t_1^2$$

hence
$$\mu_d = \frac{1}{\cos}\,[\sin\theta - \frac{2L_1}{g t_1^2}]$$
$$v_1 = a t_1 = g\,[\sin\theta - \mu_d \cos\theta]\, t_1$$

so
$$v_1 = \frac{2L_1}{t_1}$$

***Then answer the following questions***

What is the sign of $g$ in this text?

*Part* a):

What does $\theta$ mean in the two lines of the "answer" devoted to the static case? Is it:

- the angle at which the skier can get into motion?
- one of the various angles at which sliding can occur?
- one of the various angles at which sliding cannot occur?

Has the variable $\theta$ been expressed here in algebraic form?

*Parts* b), c), d):

Is the motion of the skier

- uniform?
- uniformly accelerating?
- some other case?

Whatever your answer, state to what property of friction, and/or to what assumption, this motion is due.

Luc ALPHAND (triple world downhill champion: 1995, 1996, 1997) begins his descent (under the conditions given in the text) at the same time as a brother of the same physical build (geometrically speaking), but who is much lighter.

Will they arrive at the bottom at same time

- in accordance with the model given here?
- in reality?

Discuss: under what circumstances the mass of the moving object is not involved in the equation(s) of motion (one, or several dimensions respectively)?

Now assumed to be of the same build and weight, Luc ALPHAND and his hypothetical twin begin their descent at the same time under the conditions given in the text. One of them has skis which are twice as wide as the other, but the same length. According to the model given here, will they both arrive at the same time? Whatever your answer, state to what property of friction, and/or to what assumption, this result is due, and discuss.

The validity of the expression found for the acceleration $a$ may be checked by examining the sign of this variable. From zero initial speed, the skier starts his descent and hence the value for $a$ must (at least at the start) be positive downwards, given the choice of orientation for the axis $\overrightarrow{0x}$. Does this impose some particular condition? Discuss. (consider the following experiment: if you push a piece of furniture, you may be surprised at the sudden drop in resistance the moment the furniture starts to move). (*Illustrates the fact that* $\mu_s > \mu_d$)

The text says nothing about the skier other than that he has a non-zero mass. Does the diagram in the answer sheet suggest that the skier is assumed to be a point mass? Is such an assumption required to make use of this answer sheet? If not, what does the point of intersection for the forces shown represent? Does the component of force normal to the piste exerted by the piste on the skier pass through his centre of mass, or not? Is the frictional interaction with the snow uniform over the whole area of contact? Justify your answer.

In the analysis suggested for this problem, does the coefficient $\mu_d$ depend on $\theta$? From this point of view, discuss the expression given in item d) in the answer sheet.

## 4.5 The slide projector

This topic has already been covered in Section 3.5. What is proposed here is an enhanced statement of the problem with a "bridge" between a classical and a graphical presentation.

### *Standard version*

We wish to project onto a screen at a distance $L$ a slide of height $h$ ($L$ is the distance between the screen and the slide). For this purpose we use a convergent lens of focal length $f'$.

Where should the lens be placed in order for an image to be formed on the screen? What conditions on $f'$ and $L$ will make this possible? How many potential positions are there?

Give an algebraic expression for the positions required.

Do the calculations, taking $f' = 10$ cm and $L = 4$ m. Simplify the result using the fact that $f'$ is much smaller than $L$. Remember, that when $x$ is much smaller than 1, $\sqrt{1+x} \approx 1 + x/2$. Which of the positions for the lens did you choose, and why?

### *Enhanced version*

*Parts A and B of this exercise may be done independently.*

*A - Graphical representation for optical conjugation*

Show that the usual formula for conjugation between the respective positions of the object ($p = \overline{OA}$) and its image ($p' = \overline{OA'}$) given by a lens of focal length $f'$ can be put in the form $(p' - f')(p + f') = -f'^2$. Using the change of variables $d = p + f'$ and $d' = p' - f'$, find an expression for the function $d'(d)$. Draw the function $p'(p)$ on another graph for the case $f' > 0$. Indicate the values of $p$ for which $p' = -p$. Show that if $A_1$ ($\overline{OA_1} = p_1$) and $A'_1$ ($\overline{OA'_1} = p'_1$) are conjugate, then so are $A_2$ ($\overline{OA_2} = -p'_1$) and $A'_2$ ($\overline{OA'_2} = -p_1$).

Use this curve to find the positions of the object which fulfil the following two conditions: the image is real, and larger than the object.

*B - Algebraic analysis of the slide projector*

We wish to project onto a screen at a distance $L$ a slide of height $h$ ($L$ is the distance between the screen and the slide), using a convergent lens of focal length $f'$.

Where should the lens be placed in order for an image to be formed on the screen? What condition on $f'$ and $L$ will make this possible? How many potential positions are there?

Give an algebraic expression for the positions required.

Do the calculations, taking $f' = 10$ cm et $L = 4$ m. Simplify the result using the fact that $f'$ is much smaller than $L$. Remember, that when $x$ is much smaller than 1, $\sqrt{1+x} \approx 1 + x/2$. Which of the positions for the lens did you choose, and why?

*C - Graphical interpretation*

On the graph $p'(p)$ required in part A, draw a straight line representing the function $p' = L + p$ and use this to find the answers to the three first questions in part B.

On the graph $p'(p)$ show the position $p$ found for the slide in part B and check that it is consistent with your answer to the last question in part A.

## *Elements of the answer sheet for the enhanced version*

*Part A*

Put $p = \overline{OA}$, $p' = \overline{OA'}$ and $f' = \overline{OF'}$

Regardless of the sign of $f'$, the usual conjugate point formula is:

$$\frac{1}{p'} - \frac{1}{p} = \frac{1}{f'}$$

*i.e.* $$f'(p-p') = pp'$$

so $$(p'-f')(p+f') = -f'^2 + f'(p'-p) + pp'$$

hence $$(p'-f')(p+f') = -f'^2$$

Using $$d = p+f' \quad \text{and} \quad d' = p'-f',$$

the expression $$(p'-f')(p+f') = -f'^2$$

may be written as $$dd' = -f'^2$$

The representative curve is a hyperbola.

The function $p'(p)$ may be deduced from the function $d'(d)$ by a change of variable, corresponding to a change of axes on the graph of the curve.

In the case of a convergent lens ($f' > 0$), we obtain Figure 4.7 below.

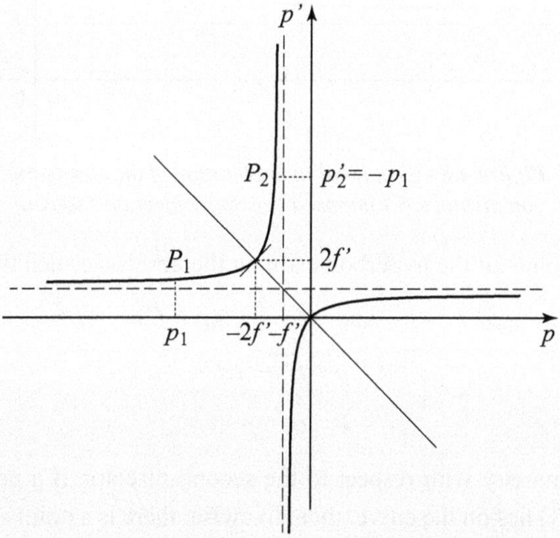


***Figure 4.7** - Curve showing object-image conjugation for a convergent lens of focal length f'*

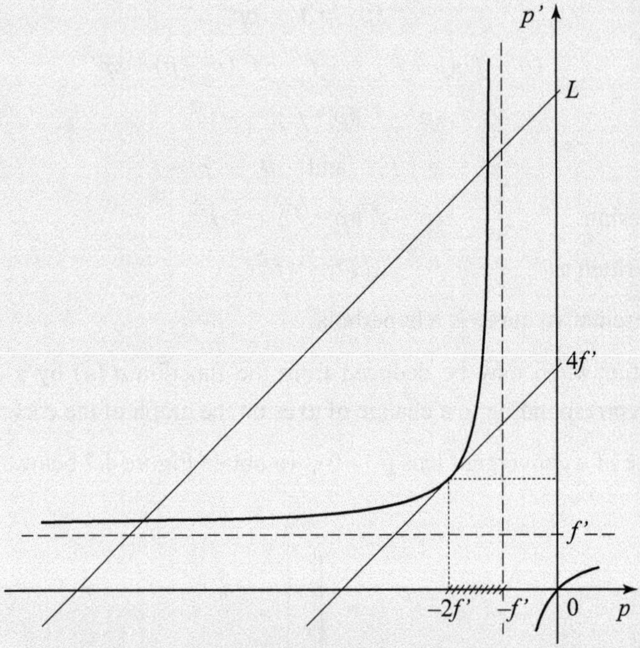


***Figure 4.8*** *- Graphical interpretation of the constraint on distance (L) imposed between object and screen*

The turning points of the hyperbolae are on the abscissae such that

$$p' = -p \quad \text{and} \quad (p'-f')(p+f') = -f'^2$$

*i.e.*:
$$p+f' = \pm f'$$

hence
$$p = 0 \quad \text{or} \quad p = -2f'$$

Given the symmetry with respect to the second bisector, if a point associated with $P_1(p_1, p'_1)$ lies on the curve, then, likewise, there is a point associated with $P_2(p_2 = -p'_1)$ and $(p'_2 = -p_1)$.

In order for the image to be real $(p < -f')$ and larger than the object $(p' > -p)$, it is necessary and sufficient that $-2f' < p < -f'$, as the graph of the conjugation curve shows.

*Part B*

The object-image distance $\overline{AA'} = \overline{AO} + \overline{OA'} = -p + p'$

is fixed
$$p' - p = L \tag{1}$$

furthermore,
$$\frac{1}{p'} - \frac{1}{p} = \frac{1}{f'} \tag{2}$$

Equations (1) and (2) $\longrightarrow p^2 + Lp + f'L = 0$

$p$ is the solution to a quadratic equation which has two roots if, and only if, $L^2 - 4f'L > 0$, *i.e.*: $L > 4f'$ (there is a double root if $L = 4f'$).

$$p = \frac{L}{2}\left(-1 \pm \sqrt{1 - \frac{4f'}{L}}\right) \quad \text{and} \quad p' = L + p$$

$$p = \frac{L}{2}(-1 \pm \sqrt{1 - 0.1})$$

So if $x << 1$, then

$$\sqrt{1+x} \approx 1 + \frac{x}{2} + \frac{x^2}{8} + \cdots \quad \text{hence} \quad \sqrt{1-0.1} \approx 1 - 0.05 - 0.001$$

*i.e.*
$$\sqrt{1-0.1} \approx 0.949,$$

which leads to:

$$p_1 \approx -3.898 \text{ m} \qquad p'_1 \approx 0.102 \text{ m}$$

$$p_2 \approx -0.102 \text{ m} \qquad p'_2 \approx 3.898 \text{ m}$$

We choose position $p_2$ since then $p'_2 > p_2$, ensuring a linear magnification ($\gamma$) whose absolute value is greater than 1 ($\gamma = \frac{p'}{p}$): the image is then greater than the slide (the object).

*Part C*

The expression $p' = L + p$ is the equation of a straight line of slope 1 and ordinate at the origin $L$. In order for an image to form on the screen, the lens should be placed at one of the positions corresponding to the abscissae of the two points where this straight line intersects with the conjugation curve (Fig. 4.8). For solutions to exist, the limiting position of such a straight line is given by the tangent at point $[-2f', 2f']$, and occurs at $L = 4f'$.

The location of abscissa $p_2$, in close vicinity to $-f'$, is well within the zone (hatched area in Fig. 4.8) such that $-2f' < p < -f'$; this represents all the conjugate points such that the object is real and that the image is greater than the object. This image is inverted.

## 4.6 Flotation between two immiscible liquids

The following exercise is commonly proposed in hydrostatics. An outline answer sheet, as shown below, enables the formal simplicity of the standard treatment to be assessed. Teacher notes are given in italics.

## Standard version

A solid cylindrical body of height $H$, cross sectional area $S$ and mean density $\rho_s$ is in a container filled with two immiscible liquids of densities $\rho_1$ and $\rho_2$, respectively, such that $\rho_1 > \rho_s > \rho_2$. Determine the equilibrium position of the cylinder, assuming that it is internally ballasted in such a way that it floats vertically.

*We may, or may not, choose one (or some) of the following comments:*

- The body floats in the first liquid ($\rho_1 > \rho_s$) and sinks in the other ($\rho_s > \rho_2$).
- The hydrostatic equilibrium existing "between two liquids" may be as per one or other of the two situations (a) or (b) shown in Figure 4.9.

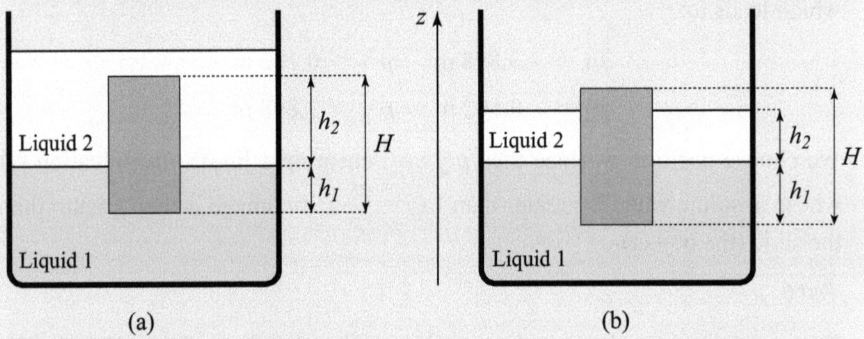


***Figure 4.9** - Two cases for a cylindrical solid at hydrostatic equilibrium "between" two immiscible liquids.* (a) *The cylinder is covered by liquid 2.* (b) *The cylinder is not covered by liquid 2. The body is then in contact with the air.*

## Outline of the answer sheet

For any position of the solid, the relationship of fluid statics $\Delta p = -\rho\, g\Delta z$ can be used, with an upward directed axis, for each part of the cylinder immersed in each liquid, of respective heights $h_1$ and $h_2$. Therefore the differences in pressure between lower and upper horizontal sections of the cylinder immersed in, respectively, liquid 1 and 2 are given by:

$$\Delta p_1 = \rho_1\, gh_1 \quad \text{and} \quad \Delta p_2 = \rho_2\, gh_2$$

The potential contribution of the air (case (b) in Fig. 4.9) can be neglected with respect to the two others, given that the density of the air is typically a thousand times smaller than those of the liquids.

The equilibrium position of the cylinder when floating is given by a Newtonian balance of forces involving the weight of the cylinder and the thrust of the liquids on each face.

The contributions of the liquids in this arrangement are calculated via the difference in the forces acting on the lower and upper horizontal faces of the cylinder, bringing into play the difference $\Delta p_T$ in the pressures at these two heights. As the atmospheric pressure on the cylinder is assumed to be constant in both cases, this difference $\Delta p_T$ is equal to the difference in pressure between the lower horizontal face immersed in liquid 1 and the upper cross-section of the part immersed in liquid 2.

$$\Delta p_T = \Delta p_1 + \Delta p_2 \quad i.e. \quad \Delta p_T = \rho_1\, g h_1 + \rho_2\, g h_2$$

The Newtonian force balance is therefore

$$(\rho_1 h_1 + \rho_2 h_2)\, Sg - \rho_s\, HS\, g = 0,$$

*i.e.*

$$\rho_1 h_1 + \rho_2 h_2 = \rho_s\, H$$

or

$$h_1 = (\rho_s\, H - \rho_2\, h_2) / \rho_1 \tag{1}$$

On account of the Newtonian force balance, we may also note that the difference $\Delta p_T$ of the pressures between the upper and lower levels of the cylinder when in hydrostatic equilibrium is

$$\Delta p_{eq} = \rho_s\, gH \tag{2}$$

These relations are in no way restricted to the particular situation involved ((a) or (b), Fig. 4.9). They merely assume that floating occurs, *i.e.* that there is enough of liquid 1 for the solid to float without coming into contact with the bottom of the container.

Where the cylinder is covered by liquid 2 we have:

$$h_1 + h_2 = H$$

Expression (1) then gives

$$h_1 = h_2\, (\rho_s - \rho_2) / (\rho_1 - \rho_s) \tag{3}$$

or

$$h_1 = H\, (\rho_s - \rho_2) / (\rho_1 - \rho_2) \tag{4}$$

In case (b), the height of the additional liquid gives the value of $h_2$, allowing expression (1) to be used directly in finding $h_1$.

Finally, expression (2) underlines the fact that the pressure $\Delta p_{eq}$ necessary for flotation to occur depends only on the mean density $\rho_s$ and height $H$ of the cylinder (for a given value of $g$).

### *Another, more disconcerting approach*

Although formally equivalent, there can be a counterintuitive interpretation of the situation:[54]

Consider the previous cylindrical solid floating in liquid 1 alone.

*Question*: if a quantity of liquid 2 is added on top of liquid 1 (covering the cylinder or otherwise, as in cases (a) and (b), Fig. 4.9), what happens to the cylinder?

1– It rises with respect to its initial position (why, and to what point?).

2– It sinks with respect to its initial position (why, and to what point?).

3– It remains in its initial position (why?).

The solution to the exercise above provides the answer. Expression (1) says that, at hydrostatic equilibrium, a high value of $h_2$ implies a low value of $h_1$. Hence, the more liquid 2 is added, the higher the cylinder. In other words, the contribution of liquid 2 to the Archimedean thrust is additive, and the higher $h_2$, the greater this contribution.

Further insights may also be drawn from the case where the added liquid is the same as liquid 1 (even though the ordered inequality $\rho_1 > \rho_s > \rho_2$ is not then satisfied): the cylinder then rises again and floats on the new surface.

The exercise might seem relatively simple and straightforward. However, the questionnaire may elicit a significant proportion of type 2 or 3 responses and considerable uncertainty.

Focusing on the weight of the added liquid may justify response 2, while the idea that it is not possible for liquid 2 to make the cylinder float might favour response 3.

In both cases, the change in the setting due to the addition of liquid 2 is not envisaged in a systemic way, but very locally. Instead of recognising that a change has occurred to the whole system (here in pressure), many students seem to consider that only a local change occurs.

This suggests the following proposition.

54 This questionnaire was inspired by a recent study: BENNHOLD C. & FELDMANN G. (2005) Instructor Notes On Conceptual Test Questions, In *Giancoli Physics - Principle with applications*, 6th edition, Pearson, Prentice Hall, 290-291. In BENNHOLD & FELDMANN's wording, it is said that "an object floats in water with ¾ of its volume submerged", and "oil is poured on top of the water"; and the questions are slightly different. In the sample concerned (1st year university, unspecified number of students), 75% of students gave responses of type 2 or 3. See also VIENNOT L. (2011) *Floating between two liquids* (www.eps.org, select *Education* and then select *MUSE*).

## *Make use of the graphs for a better grasp of the physical situation*

Hydrostatic questions raise the crucial issue of pressure gradient. If the dependence of pressure $p$ on height $z$ (measured from the bottom of the container) is represented graphically, then this aspect can be visualised via the slope of the straight line representing the function $p(z)$.

Students can thus be encouraged to draw a graph such as in Figure 4.10, representing the pressure as a function of height with (black curve) or without (dotted curve) the second liquid.

The relationship between such a graph and the physical situation may not be grasped immediately (far from it, in fact) so it's worth spending a little time on it. Working along these lines, it may well be profitable to consider a representation which although less familiar than an algebraic calculation, has the advantage of eliminating any exclusively local analytical difficulties and focussing on what is essential. And, as a spinoff, it seems reasonable to expect that activity of this type will, in the long run, contribute to a better grasp of the linear relationship in physics via a graphical representation.

First of all, it is perhaps worthwhile looking at the following questions:

- The physical meaning of a horizontal straight line?
- The meaning of the slope of a straight line?
- The meaning of two parallel segments of a straight line?
- What happens to the graph if the atmospheric pressure rises?
- Considering an order of magnitude for the atmospheric pressure as compared with the pressure differences in this situation, does this allow the origin of the pressures to be located at the intersection of the axes?

Once these issues have been clarified, the actual work phase can begin.

### *Comments to be discussed with the students:*

- Adding liquid 2 alters the whole scalar pressure field in the container. *This point is essential to initiate a line of reasoning aimed at the restrictions of a local analysis.*
- There is a graphical equivalent to the approximation made above for the role of the air (its negligible contribution to the Archimedean upthrust: a constant value of atmospheric pressure around the container).

- What counts in assessing the fluid upthrust on the cylinder is the pressure difference $\Delta p_T$ at the lower and upper faces of the cylinder, *i.e.* at two heights which differ by $H$ (the height of the cylinder). Its equilibrium value $\Delta p_{eq}$ is determined by the object itself (its weight and cross-sectional area).
- Making a diagnosis on some arbitrary position of the cylinder (in equilibrium or otherwise, upward or downward resultant of forces) is essentially that of seeing how the perpendicular segments (characteristic of the cylinder) representing the values of $\Delta p_{eq}$ and $H$ fit the graph (a kind of "graphical set-square") which is characteristic of the fluid medium. The mental, or even physical, manipulation of this "graphical setsquare" against the slopes of the straight lines representing the pressure field is symbolically equivalent to a mechanical adaptation of the object to the fluid medium. Figure 4.11 shows various possible situations.

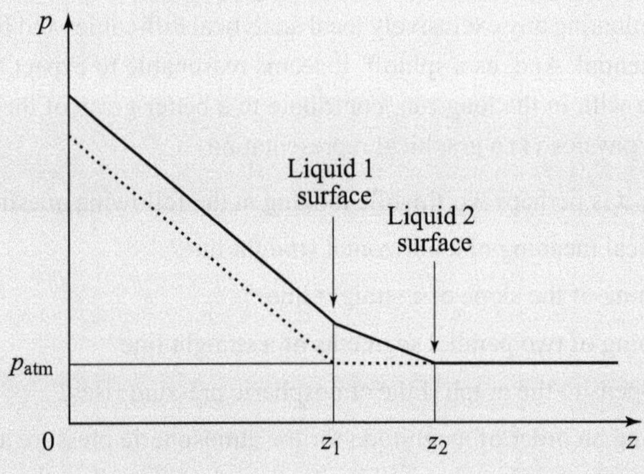


***Figure 4.10** - Pressure versus altitude (z: origin at the bottom of the container) in two situations: only liquid1 (dotted line), liquid 2 added on top of liquid 1 (solid line)*

## Experimentation

*Some extremely simple equipment (a plastic cup or glass, a tube of pills or a "Kinder egg", ballast, oil and water) is all that is needed to illustrate all the cases in the figure, in conjunction with the algebraic and graphical analyses given here.*

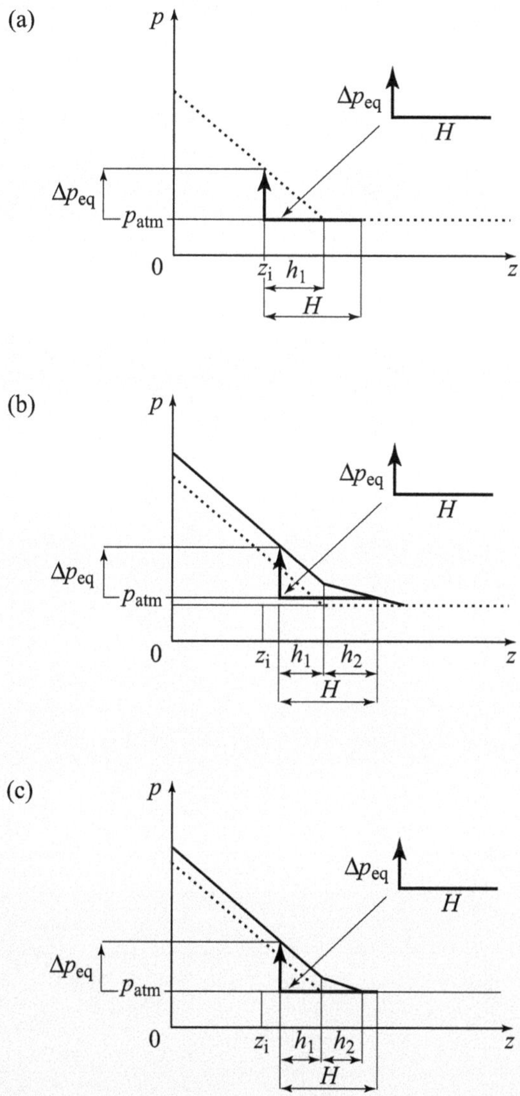


***Figure 4.11*** - *Use of graphs p(z) for adjusting the "setsquare" representing $\Delta p_{eq}$ vertically and H horizontally.*

(a) *With liquid 1 alone, the cylinder floats above the waterline.*

(b) *Copious quantities of liquid 2 are added. Since $\rho_s > \rho_2$, the cylinder does not break the surface of the liquid into the air. Nevertheless, the lower surface rises with respect to position $z_i$ in case* (a) *($h_1$ is smaller than in case* (a)).

(c) *With just a small amount of liquid 2 added, the cylinder again rises to the surface ($h_2$ is equal to the thickness of liquid 2). Again, $h_1$ is smaller than in case* (a), *and hence the cylinder rises.*

# Part II

## PHYSICS: LINKING FACTORS

# Chapter 5

# Links between Phenomena in Terms of Type of Functional Dependence

## 5.1 Introduction

If we are to discuss the cost-benefit ratio for the functional approach and its graphical translation, we should not limit ourselves to one particular physical situation. Besides the elegance of the solutions it offers, this approach has the advantage of facilitating access to a remarkable aspect of physics–without prejudice to the other sciences: there are few laws which encompass a multitude of phenomena[55]. Of course, it's important in teaching that the accent should be laid on such a major asset, without letting the supposed evidence cause it to lose its power. Indeed, there is an extra linking factor in that which brings together phenomena governed by functional dependencies of the same mathematical form. Thus it is with the model of the harmonic oscillator, a central theme of the 1995[56] *Terminale* (Year 13) programme, still very much in evidence at this level the next year,[57]and a favoured topic for the introduction of waves.[58] Similarly, the theoretical power of the Poisson distribution,

---

55 See, in France, the 1992 Year 9 programme, BO 31 of 30 July 1992, p. 2088: "The teaching process must make it clear that physics is an essential cultural element, by showing that the world can be understood and that the extraordinary richness of nature and technology can be described using a small number of universal laws of physics making up a coherent picture of the universe". The word 'parsimony' is currently used to get this idea across. The theme of links, so essential for understanding, is thus at the heart of the conceptual construct: "And yet, a person understands some information available to him or her only if he or she grasps the connections, the relationships, between phenomena, concepts and ideas to which the information refers. It can be said that the understanding of information consists precisely in the grasping of such relations." Igor Kluvánek, cited by R. Nillsen (2009) Can the love of learning be taught? *The Pantaneto forum*, Issue 36 (http://www.pantaneto.co.uk/issue36/nillsen.htm).

56 MEN (1995) *Bulletin Officiel de l'éducation Nationale* du 16/2/1995, *Terminale* (Year 13) science programme.

57 MEN (2001) *Bulletin Officiel de l'Education Nationale*, Special edition 30 August 2001. Secondary school programme, *Cycle Terminale* (Year 13). Applicable to start of school year 2002.

58 e.g. Crawford F.S. (1965) *Berkeley Physics Course,* **3**, *Waves*, McGraw-Hill Company, New York.

the decreasing exponential, or the BOLTZMANN factor are the stuff of dreams for any physics teacher.

Simplicity (kept intentionally) suggests an example using the most accessible of functions, linear dependence: $y = ax$ ($a$ being a coefficient independent of variables $x$ and $y$). Here, this will be $d = vt$, a relationship accompanied by various changes of origin for the variables. In words, this means that the distance covered at constant speed increases in the same ratio as the duration of the event. Here, we are once more back again with problems involving trains passing each other.

Why not? although, admittedly, restricting ourselves in this way isn't terribly exciting.

To illustrate an alternative approach, we shall explicitly mention (a little more than before) some recent research work. Let us restate the underlying idea. The *principle* of teaching practice involved here is not new; rather we are only talking of putting something into *practice* in realistic terms, in acceptable amounts, with an unusual and/or surprising example of relative simplicity, and the effect it produces on various audiences.

## 5.2 Delayed signals: from stars to bats

*In collaboration with Ivan FELLER*

The Year 11 programme which was introduced in 2000[59] in France mentions the topic of delayed signals, illustrated on a cosmic scale: since light does not travel at infinite speed, the light from the stars reaches us after a period of time. Light carrying information from the moon takes a little over one second, while for the sun it is about 8 minutes; for the stars and galaxies the time required soon mounts to millions of years, or even several orders of magnitude more. In the accompanying document, we find a "problem-situation" intended to draw attention to this point (Fig. 5.1). Two present-day extra-terrestrials are contemplating the attempts of some hairy Earthlings to make fire. The question one of the extra-terrestrials asks his companion is put to the student: at what distance from the Earth must the observing planet be? If the terrestrial scene were to be dated say, three hundred thousand years before the present, it would follow that the distance would be three hundred thousand light-years.

This is a document which certainly strikes the imagination, for the drawing clearly shows the various continents on a pretty blue background (the Earth has to be recognisable). It's hard to imagine that, seen from such a distance, the Earth would appear as anything other than a point, not to mention the problem of its luminosity.

59 Document accompanying the programme of physics of *Seconde* (Year 11 programme) © MENRT, CNDP and GTD of physics-chemistry.

**Activity B2 - An image...delayed!**

This image can be copied onto a transparency and projected onto a screen.

It presents a problem-situation around which a group discussion can be organised in class.

The scene takes place in the year 2000.
Can you answer this extra-terrestrial's question?

***Figure 5.1*** - *A suggested document to accompany the Year 11 programme (French programme of Seconde) (programme 2000)*

This story illustrates the road to educational hell, paved as it is with good illustrational intentions (already commented on above); it could be useful to exploit this in a positive way. One way of doing this is to conduct a critical analysis of this document with the students, in particular from the perspective of the constraints of popularised illustration.[60]

60 Recent research has assessed this type of teaching strategy: see Ivan Feller, thesis 2008. *Usage scolaire de documents d'origine non scolaire : éléments pour un état des lieux et étude d'impact d'un accompagnement ciblé*. Université Paris-Diderot (Paris 7). Online: halshs.archives-ouvertes.fr/docs/00/36/63/.../These_Ivan_FELLER_op.pdf. See also Appendix F and Feller I., Colin P. & Viennot L. (2009) Critical analysis of popularisation documents in the physics classroom. An action-research in grade 10, *PEC*. **17** (17) 72-96.

Let us reconsider the topic of the links between phenomena. First year students were invited to find a familiar phenomenon falling within the same formal description. Sometimes we get this very pertinent reply, namely the difference between seeing a lightning flash and hearing the thunderclap which follows. In order to show the curves (actually straight lines) representing the propagation of the various signals involved it can be helpful to draw a graph (Fig. 5.2). The slopes of these straight lines give the speed of each signal, with the scale of the graphs appropriate to the phenomena. Hence for interplanetary signals, it is important that the slope of the straight line should not appear infinite; in the lightning case on the other hand, it emphasises that the light arrives virtually instantaneously as compared with the sound.

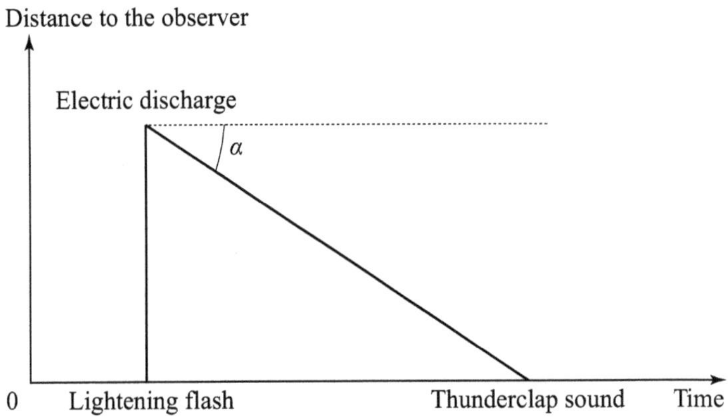


***Figure 5.2*** - *Type of graph showing how an atmospheric event is perceived: the flash of light is seen before the thunderclap is heard; the slope tan α of the straight line representing the advance of the sound is negative, with a value of approximately 330 m.s*$^{-1}$. *The axes should be graduated appropriately.*

Work organised on this basis is simple without being immediate, accessible if due attention is paid, and highly relevant in the matter of training. In varied contexts and over different timescales we see the same formalism at work: we experience the unifying value of a physical theory. While, at the same time indispensable familiarity is gained with the mathematical tool involved.

The arguments often developed[61] for the harmonic oscillator or the decreasing exponential are illustrated here with an even simpler formalism: It would be a serious error to deprive ourselves of any potential benefits. In any case, it seems inappropriate to confuse "much simpler" with "too simple to be of any value".

Once the space-time graphs discussed here have been grasped, we are then armed to deal with echoes, submarine sonar, bats and moths. We can understand from this last

61 Ages 16 and older.

example why bats have a minimum and a maximum detection distance (see Fig. 5.3 and caption).

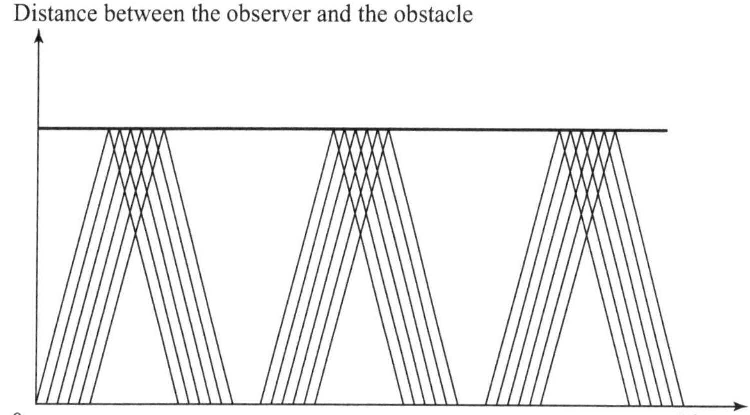


***Figure 5.3** - Type of graph representing a chirp signal with echoes from a fixed obstacle (bats, moths). For an obstacle which is too close, the outgoing and incoming signals for the same chirp will be merged. For an obstacle which is too far away, the outgoing and incoming signals will be merged for two different chirps. The axes should be graduated appropriately for the particular phenomenon. (RUMELHARD G., private communication)*

## 5.3 Graphical version of the DOPPLER effect

*In collaboration with Jean-Luc LEROY-BURY*

The unifying style of our exercise can be extended without much difficulty.

Let's take the DOPPLER effect, which has to do with propagating signals and hence a good candidate for including among our first examples.

What is it?

Remember that we have one periodic signal (of period $T_S$), its source (S), a receiver (R), and a speed of propagation $c$ (phase velocity for a wave, denoted by **c** for light *in vacuo*). In the event of relative motion between the receiver and the source (of relative speed $v_R$), the period $T_R$ at the receiver is given by the expression $\frac{T_R - T_S}{T} = \frac{v_R}{c}$. Depending on the particular case,[62] the period $T$ in the denominator is identified with $T_R$ or $T_S$.

62 For a source moving with respect to the medium ($v_S$), stationary receiver, $(T_R - T_S)/T_S = v_S/c$; for a moving receiver ($v_R$), with respect to the medium, stationary source $(T_R - T_S)/T_R = v_R/c$. The slightly more complicated case of the relativistic DOPPLER effect (*i.e.* the propagation of light *in vacuo*) is not dealt with here. For low relative speeds between source and receiver, all the expressions are the same to first order.

Apart from the classic diagrams of wavefronts getting closer or further apart with the motion of the source, or a moving observer encountering them with greater or lesser frequency, how do we show it?

In one dimension,[63] the phenomenon is frequently represented using objects deposited by a source onto some transporting device (corks in a river, ink marks on a moving belt, etc.) which advances with respect to the medium at the speed of propagation of the signal, $c$.

The source can move at speed $v_S$ with respect to this same medium. This model is easily adaptable to the case where the receiver, or even the source and receiver, is/are moving with respect to the medium.

Figure 5.4 summarises this elementary model, and also the simple calculation required to obtain the appropriate formula in the case of a receiver moving away from a source.

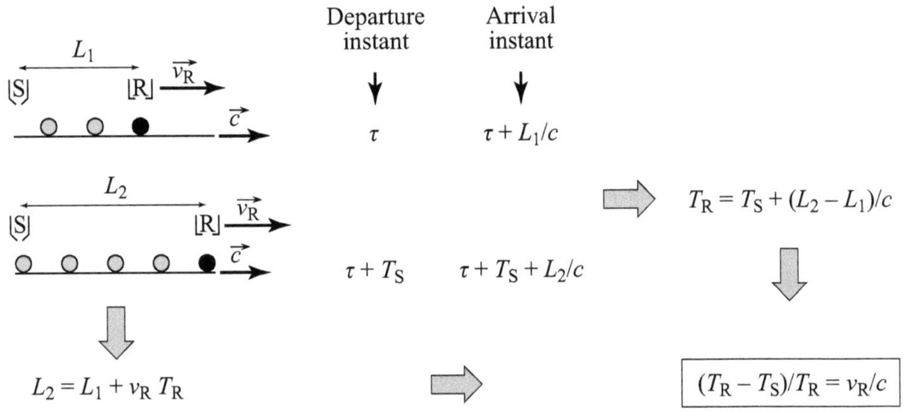


***Figure 5.4** - Example of model and elementary calculations to illustrate the DOPPLER effect. Marks are deposited on a moving belt (travelling at speed c with respect to the medium) by a source S (Period $T_S$, stationary with respect to the medium). The marks are intercepted by a receiver R (of period $T_R$) which is moving with respect to the source at constant speed $\vec{v}_R$.*

63 The whole of the end of this chapter is a reprise of a collaborative study with J. L. LEROY BURY. In particular, the figures are taken from one or other of the following publications: LEROY-BURY J.L. & VIENNOT L. (2003) DOPPLER et RÖMER : physique et mathématique à l'œuvre, *Bulletin de l'Union des Physiciens*, **859**, 595-1611 ; VIENNOT L. & LEROY J.L. (2004) DOPPLER and RÖMER: what do they have in common? *Physics Education*, **39** (3) 273-280. See also VIENNOT L. (2004) *The design of teaching sequences in physics. Can research inform practice? DOPPLER and RÖMER*, In REDISH E.F. & VICENTINI M. (Eds. Research on Physics Education. Course CLVI, SIF Varenna. Amsterdam: IOS press, 521-532.

What does this have to do with the business of our linear relationship, apart from the formula $d = vt$ clearly at work in this calculation? And likewise, where is the problem, if one exists?

Indeed, there is one. Many students at various levels believe (either before or after being taught) that the shift in period between the signals emitted and received is due to the relative velocity between receiver and source, witness the "yeeeeoooww" noises of Formula 1, as well as more formal investigations.[64] However, given this relative speed, their answers also imply that distance is important, even though this particular variable does not feature in the expression for the DOPPLER effect.

Each time distance and relative speed are coupled, they do in fact have good reasons for thinking this. Such is the case for radial velocity (projected onto the line joining source and receiver) and the distance to the moving object when, for example, a Formula 1 car goes past the stands where the observer is a good distance away.

This is especially the case for the galactic redshift associated with the expansion of the universe, a favourite topic in the media. The issue is really not one of making the process incomprehensible to the students,[65] but rather of helping them make sense of it all.

Hence we might want to put the accent on the fact that, for the DOPPLER effect, the crux of the matter is the dependence on the relative speed between source and receiver.[66]

This concern can be approached from the perspective illustrated at the start of this chapter: capitalise on the graphs expressing the one-dimensional propagation of signals, here the intersection of the wavefronts with the line joining source and receiver. From the graphical representation of interception by a receiver at a fixed distance from the source (Fig. 5.5), and also relevant in analysing the echo from a fixed object (Fig. 5.3), it is natural to see what happens if the obstacle or the receiver were to move, at constant speed, with respect to the source S (Fig. 5.6). The regularly spaced oblique lines now cross the oblique line representing the motion of the receiver. As can be seen on the graph, the temporal displacements are different from those obser-

64 References in previous note.

65 It is especially understandable that these students might be lost when shown "a wavelength" leaving the source on the same figure (in the same reference frame?) if we interpret a little squiggle drawn on the source and another, different squiggle arriving at the receiver, for which the signal appears to come into being before arriving. See, for example, BOTTINELLI L., BRAHIC A., GOUGUENHEIM L., RIPERT J. & SERT J. (1993) *La Terre et l'Univers* [The Earth and the Universe], Col. Synapses, Hachette Éducation, Paris. *L'effet DOPPLER-FIZEAU* [The DOPPLER-FIZEAU effect], box p. 137; or FRANÇON M. (1986) *L'optique moderne et ses développements depuis l'apparition du laser* [Modern optics and developments since the laser], Col. Liaisons scientifiques, Hachette-CNRS, Paris, p. 74.

66 For simplicity we make use of the most general case where one or other of these elements is stationary with respect to the medium (if there is a medium) or the case of light in a vacuum.

ved for the fixed receiver. In other words, the receive period is different, longer when the receiver is moving away, and shorter when approaching.

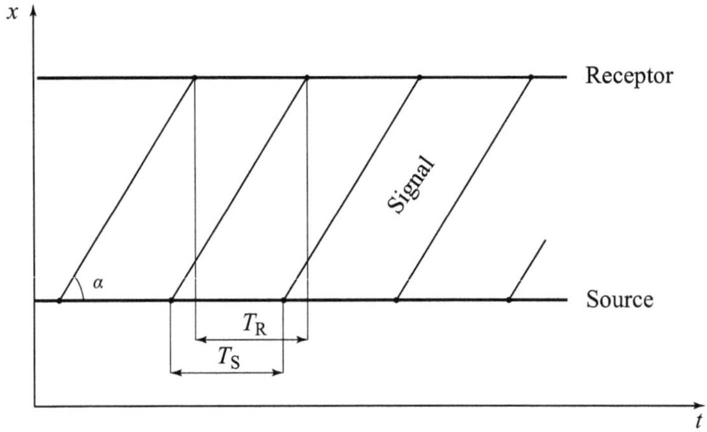


***Figure 5.5*** *- Temporal graph for the propagation of signals emitted at regular intervals by a stationary periodic source with respect to the medium, and received by an observer who is also stationary with respect to the medium: the receive period* $T_R$ *is equal to the emitted period* $T_S$.

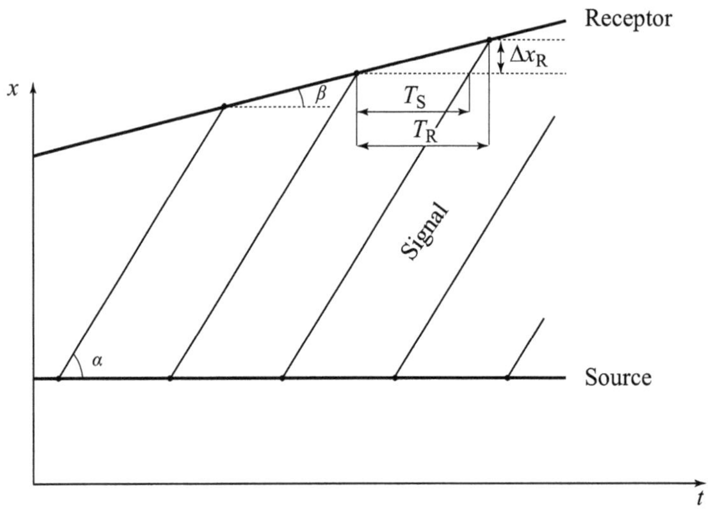


***Figure 5.6*** *- Temporal graph of the displacement of signals emitted at regular intervals by an observer moving away at constant speed v with respect to the medium: the receive period* $T_R$ *is longer than the emitted period* $T_S$. *The displacement of the observer,* $\Delta x_R$, *during the period* $T_R$ *is expressed in two ways:* $\Delta x_R = vT_R = c(T_R - T_S)$ *i.e.* $(T_R - T_S) / T_R = v/c$.

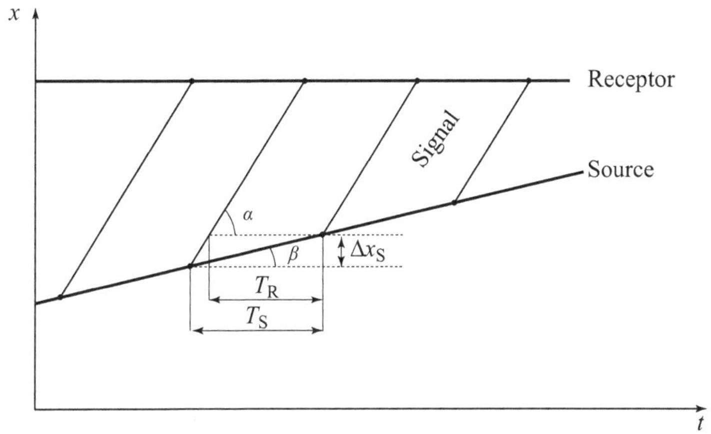


***Figure 5.7*** *- Temporal graph of the displacement of signals received by an observer stationary with respect to the medium, and emitted at regular intervals by source moving at constant speed* $v_S$*: the receive period* $T_R$ *is shorter than the emission period* $T_S$*. Displacement of the source* $\Delta x_S$ *during the period T is expressed in two ways* $\Delta x_S = v_S\, T_S = c\,(T_S - T_R)$*, if we put* $u = - v_S$*, to express a receding speed (here negative), then* $(T_R - T_S) / T_S = u/c$

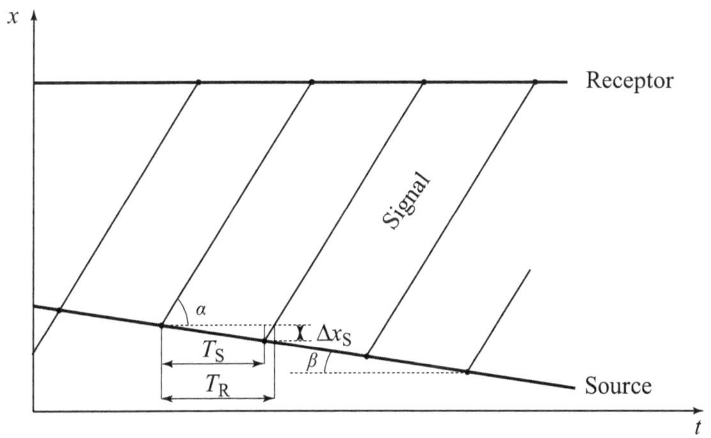


***Figure 5.8*** *- Temporal graph for a situation analogous to that dealt with in Figure 5.6, with the source this time moving away from the observer* ($v_S < 0$)*: the receive period* $T_R$ *is longer than the emission period* $T_S$*. As for the situation in Figure 5.7, the displacement* $\Delta x_S$ *of the source during period* $T_S$ *is expressed in two ways:* $\Delta x_S = v_S\, T_S = c\,(T_S - T_R)$*; if we put* $u = - v_S$ *to express a receding speed (here positive), then* $(T_R - T_S)/ T_S = u/c$.

We can see clearly that it's a question of slopes, and that any straight line displaced parallel to itself (distance variation) does not change the receive period. That means that when the teacher emits a pure sound, the student hears the same note at the back of the class as at the front, but anyone moving rapidly would hear a different note.

Does this mean that we are stuck at a rather vague qualitative level? Can we find the relationship involved? The answer lies in considering two slopes and the associated right triangles. As the graph (Fig. 5.6) suggests, the variation in the source-receiver distance $\Delta x_R$, between two receive events can be calculated in each case: $c\ (T_R - T_S) = v_R.T_R$. Only the time and slopes and a shift between two positions are involved in this calculation; this is not at all the case with the overall distances covered by the signal. The accent is on the variational aspect.

This is not, therefore, just a slightly artificial exercise in mental gymnastics intended to improve familiarity with graphs. These can really be put to good use in a concise analysis focused on the essential features. Incidentally, it might be added, if practice with graphs and understanding the meaning of the slopes are corroborated by the exercise, then we can take some pleasure in this.

## 5.4 Even more links? DOPPLER and RÖMER

Nearly two hundred years before DOPPLER, the Danish astronomer Ole Christensen RÖMER encountered a relative displacement effect between source and receiver (1676). The source was one of Jupiter's satellites, Io for example, which, when not hidden by Jupiter, scattered the light it received from the Sun in the direction of the Earth. The receiver was an observer behind a telescope in the Paris observatory. The relative displacement was mainly[67] due to the orbital motion of the Earth. Observing the lags or advances in the emergence of Jupiter's satellites according to the season, RÖMER concluded that the speed of light was finite, and he provided a good order of magnitude for it. His demonstration involved the cumulative effects of displacement over the entire diameter of the Earth's orbit. However, we can give an equivalent formulation which emphasises that this is a DOPPLER effect ahead of its time (or at least before its official discoverer).

Let's resume the story above. The source is Io, which makes Jupiter into a kind of rotating beacon with a period of 42.5 hours. The receiver is an observer on Earth, while the relative displacement is that associated with the orbital motions of the Earth and Jupiter. If the latter were fixed with respect to the sun, its distance from Earth would change from season to season as a constant to which is added the projection of the Sun-Earth radius vector onto the Jupiter-Sun axis; in other words, an harmonic function. This is nearly what we see. Figure 5.9 reproduces the corresponding position data, which is far from being a straight line.

67 Jupiter also moves with respect to the Sun.

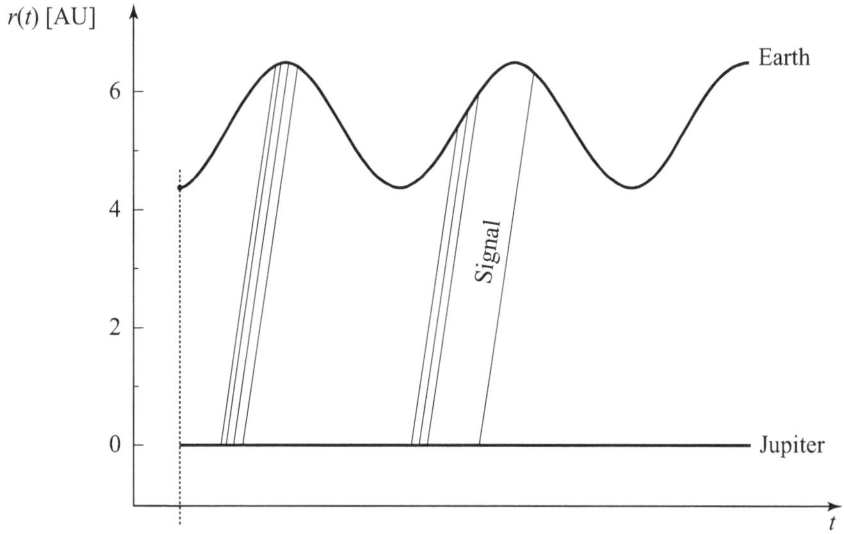


***Figure 5.9*** - *Temporal graph of the motion of the Earth with respect to Jupiter (in a jovian reference frame; AU: Astronomical Unit, i.e. the mean distance of the Earth from the Sun). A few oblique lines represent the propagation of periodic light signals from Jupiter to Earth (references in note 63).*

What happens in physics when a curve has the temerity not to be straight line? We take a little piece of it (technically speaking, we make a "local linear approximation" of the function), and analyse using this straight line segment. In our case, if we are in the vicinity of a maximum or minimum in the distance between planets, everything happens as if the receiver were at a constant distance from the source, hence no DOPPLER effect (Fig. 5.10). On the edges of the orbit as seen from Jupiter, the situation is different: the relative speed is a maximum in one direction or the other, and the DOPPLER effect is guaranteed (Fig. 5.11). From one coupling between position and distance we pass to an "anti-coupling". For distances furthest from the mean value, *i.e.* from the Jupiter-Sun distance, there is no DOPPLER shift. For two positions corresponding to this same distance, the shifts take their maximal values.

In addition, these graphs provide a good opportunity to appreciate the main significance of RÖMER's discovery. We need only ask ourselves what happens in Figure 5.11 when we assume the propagation speed of the signal to be infinite. The lines representing the propagation then straighten up to become vertical; hence there would be no difference between the receive and emit periods (Fig. 5.12). RÖMER therefore showed that the speed of light was not infinite.

Thus, provided we make the necessary effort to benefit from it, astronomy can come to the aid of teaching.

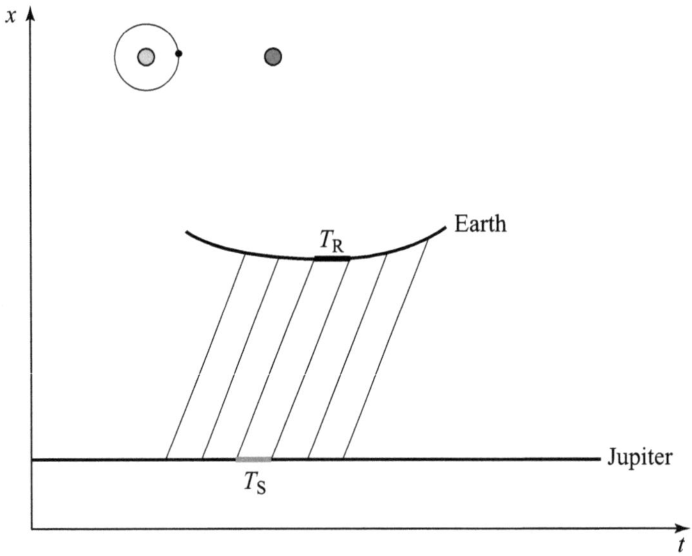


***Figure 5.10*** - *When the distance between Earth and Jupiter is a minimum (or maximum), no* DOPPLER *shift is observed in the periodic emergence of one of Jupiter's satellites*

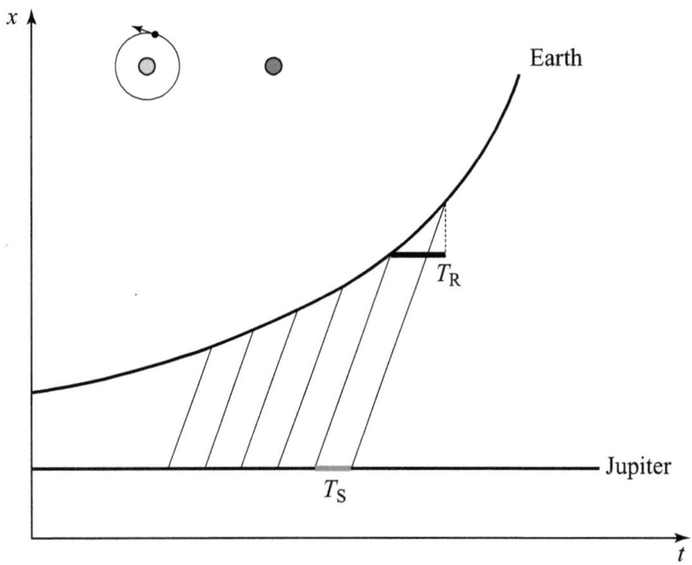


***Figure 5.11*** - *When the distance between Earth and Jupiter is some intermediate value, a* DOPPLER *shift is observed in the periodic emergence of one of Jupiter's satellites.*

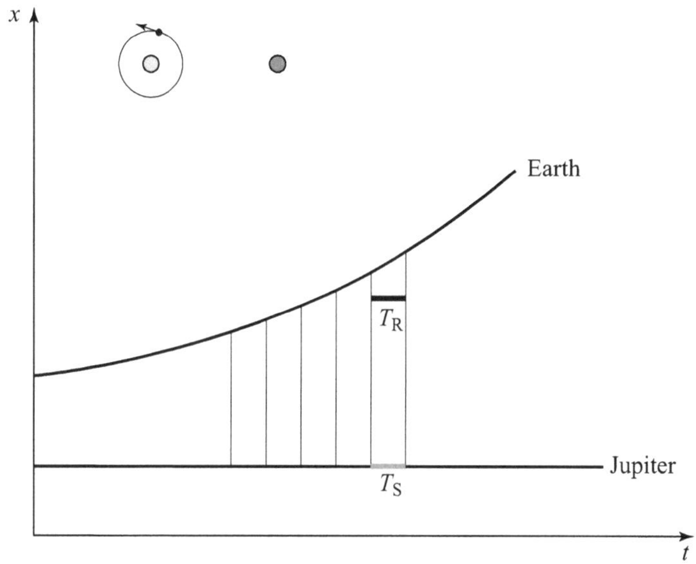


***Figure 5.12*** - *If the speed of light were infinite there would be no* DOPPLER *effect, regardless of the relative speed between Earth and Jupiter*

## 5.5 Investment?

Via the use of graphs familiar to teachers of relativity and the often associated "space-time curves", this exercise has taken us from a simple thunderclap to the DOPPLER effect and RÖMER's discovery. However, all that was needed was this relationship between the distance covered, its duration, and constant speed.

The more reticent would say that, regarding simplicity, we could do somewhat better. Indeed, the calculations are either nonexistent or elementary, but you do have to be able to read a graph. We said further back that once a variable is no longer a spatial variable, the difficulty goes up a notch. Mathematics teachers faced with this kind of analysis would agree:[68] it is hard for them, but in this respect they are not alone.[69]

In terms of abstraction, there is a real price to be paid. The analysis presented here aims to illustrate this idea: a profitable investment. Just one relationship and there you are! A whole flood of phenomena becomes comprehensible. And, as a bonus, one of the fundamental properties of physics is highlighted, namely compactness.

68 Intervention by LEROY-BURY and VIENNOT in the "Modelling" training session at IREM Paris, 2003.

69 This difficulty is also manifest among physicists: in the IUFM training session (two groups of about thirty second year trainees) and in the "physical sciences" degree (two groups of about twenty third year students); reference in note 63.

Furthermore, if that's difficult, then it hardly makes sense not to prepare, thanks to the help of a simple thunderclap or a few bats, what we then go on to use when teaching relativity and its famous "space-time curves". After all, what's the point of introducing these so abruptly, like some mundane tool, to amazed first year university students we had been desperately trying to "spare" up to that point?

# Chapter 6

## THE RELATIONSHIP BETWEEN DIFFERENT APPROACHES TO THE SAME PHENOMENON

Under the heading of similar relationships between variables, the previous chapter reconciled some phenomena which appeared at first sight to be different. In a complementary manner, this chapter will go into the theme of links: here we are dealing with a single phenomenon, or context at least, bringing our thoughts together. This will involve different approaches, especially in terms of the scale of the description used: macroscopic, mesoscopic, or even particle-based. Once again, the example will be taken from everyday life and the physics simple.

### 6.1 An instructional hot-air balloon

With a touch of irony, we can define this "instructional hot-air balloon". For such a balloon, the envelope open at the base defines an internal space of volume $V$, within which the air is at temperature $T_{\text{int}}$ and pressure $p_{\text{int}}$. The whole thing, including passengers, has mass $m_{\text{t}}$. We should simplify, and temporarily forget, for example, the turbulence generated by the burners. Initially, the results will not suffer too greatly, and much will remain understandable. The outside must also be defined: air at atmospheric pressure ($p_{\text{ext}} = p_0$) and at temperature $T_{\text{ext}}$. Very frequently[70], equality of internal and external pressures is added to the model ($p_{\text{int}} = p_{\text{ext}} = p_0$), the rationale being that the envelope is open.

A standard solution relies on ARCHIMEDES' principle: the upthrust due to the outside air on the whole ensemble is balanced by the weight of the volume $V$ of the external air.[71] This weight is yet to be evaluated, as is that of the internal air to balance the forces justifying the equilibrium required. The weights in question, corresponding to the same volume, are subsequently differentiated by different values for the density

70 By way of example: GIANCOLI D.C. (2005) Physics (6th edition): "Instructor Resource Center" CD-ROM, *Prentice Hall.*

71 With respect to this value, neglecting that of the volume of the materials of the gondola and the suspension cords.

of air, itself related to the variables already introduced by the barely transformed ideal gas law: $\rho = \frac{Mp}{RT}$. In three or four lines of working facilitated by the equality of the pressure terms, temperatures (via their reciprocals) and the problem data can all be linked together.[72] We are then in a position to know to what temperature the internal air must be heated to achieve lift-off, and subsequent stability once in the air.

## 6.2 Ritual: a pact with inconsistency?

Should we be worried about the simplifications inherent in the common approach to this classically presented problem, in particular to first-year university students?

Yes, we should indeed. As we know, physics starts off by thinking of simplifications. However, here we encounter an assumption which, taken at face value, would send the balloon crashing to earth quicker than you could say 'ARCHIMEDES'. If the pressures were the same at every point ("atmospheric pressure"), the resultant of the pressure forces acting on the envelope due to the gases present would be zero. Each part of the envelope would be subjected to two exactly opposite forces. Plus, no particular spatial direction would be preferred by these gases: why should they push upwards? Yet again, using ARCHIMEDES' principle is to make use of the sine qua non of its relevance, namely the existence of pressure gradients, essential for hydrostatic problems where gravity is present. Between the level of the opening and that at the top of the balloon, the pressure of the outside air falls. Ditto for the air inside. However, as this is less dense, the pressure from bottom to top falls less quickly on the inside than on the outside. Starting from a value assumed identical at the level of the opening, the internal and external pressures are not equal elsewhere, in particular at the top of the balloon: the highest pressure is on the inside. We then begin to grasp that the envelope can be inflated and held airborne despite the weight of the whole thing. This analysis is summarised in Figure 6.1 and is illustrated by a strange balloon, cylindrical for reasons of formal economy: there is no need for a complicated integral to show (or even to formally verify) that ARCHIMEDES' principle is consistent with a local analysis of the forces acting on the envelope. The global

72 For a balloon of total mass $m_c$ (for the solid parts), taking account of the density $\rho$ of an ideal gas of (mean) molar mass $M$, $\rho = \frac{Mp}{RT}$, and from ARCHIMEDES' principle, the Newtonian equilibrium is written: $m_c + \frac{M}{R}\frac{p_{\text{int}}}{T_{\text{int}}}V = \frac{M}{R}\frac{p_{\text{ext}}}{T_{\text{ext}}}V$, *i.e.*, assuming that the (mean) internal and external pressures are very close to their value $p_0$ at the opening, $[1/T_{\text{ext}} - 1/T_{\text{int}}] = m_c R/(p_0 M V)$.

analysis supported by the gradient theorem[73] and its consequence in hydrostatics (the expression for the ARCHIMEDES' interaction) unites the mechanical (local and more direct) balance of forces in play. Two approaches, each casting light on the other, compete for an understanding of the phenomenon. For many students (we will return to this) this was an opportunity to get to the nub of ARCHIMEDES' principle.

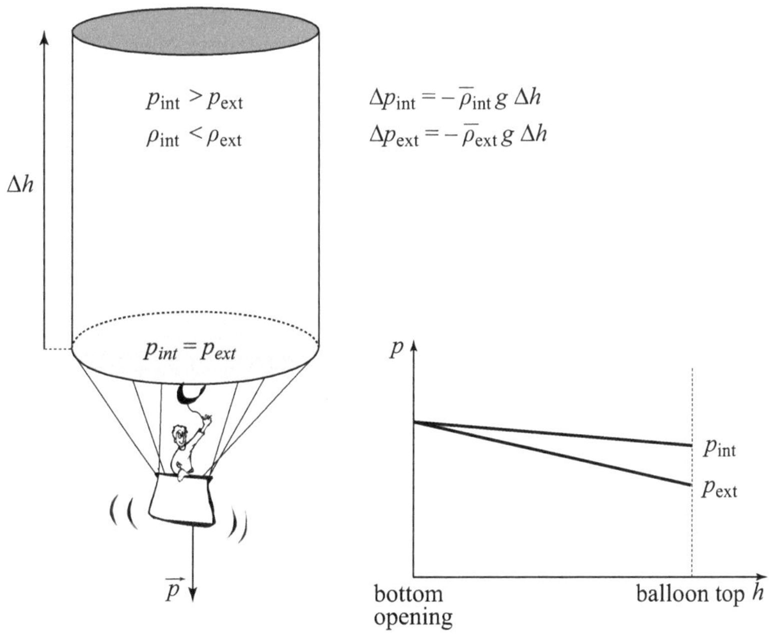


***Figure 6.1*** *- Elements for understanding how a balloon is held airborne, here shown as a cylinder to facilitate understanding the effect of pressure forces on the envelope (see the text and note 73).*

## 6.3 Two approaches for a single phenomenon

Let us pause here to consider the unusual nature of this analysis.

Several studies agree as to the reaction of individuals consulted on an exercise containing the assumption in question, namely, that "the pressure is everywhere the same".

73 The gradient theorem, applicable to a closed surface $S$ enclosing a volume $V$, and (here) to a scalar field $p$: $\iint_S p\,\overrightarrow{dS} = \iiint_V \overrightarrow{grad\,p}\; dV$ ; in a fluid of density $\rho$ at equilibrium we have $\overrightarrow{grad\,p} = \rho\vec{g}$. ARCHIMEDES' principle follows immediately.
This principle leads to the relation $[1/T_{ext} - 1/T_{int}] = m_c\, R/(p_0 MV)$ (see previous note).
Another approach, here using a cylindrical balloon of height $\Delta h$; to first order we have at the upper level: $p_{ext} \approx p_0 - \bar{\rho}_{ext}\, g\Delta h$ *et* $p_{int} \approx p_0 - \bar{\rho}_{int}\, g\Delta h$. The supporting force which acts on the upper horizontal face of area $S$, balances the weight of the solid parts if, and only if, $m_c\, g = (p_{int} - p_{ext})\, S$, which leads to the same expression as that produced by the global treatment $[1/T_{ext} - 1/T_{int}] = m_c\, R/(p_0 MV)$.

When invited to clarify or fill in the text, almost all these individuals consciously remember what a perfect gas is. There was not one who pointed out the absurdity of assuming the pressure to be uniform or identical on either side of the fabric wall, nor did anyone suggest this more realistic formulation: that the average pressures, internal and external, are very similar.

What kinds of individual are we talking about? First year (N=15) or third year (N=50) university students, the latter aiming to become teachers (N=36) or journalists and scientific writers (N=14).[74] Of course, we might say that "the students don't have a critical sense, they are only interested in applying formulae". So let's look at the teachers. Apart from those who write books[75] when invited to improve the wording of a prototype text, trainees at teacher training colleges registered en masse their hesitations about the perfect gas but said nothing about the pressures being equal: all of them[76] put up with the latent inconsistency. Is there any of us who can swear to never having shown such a blindness?

When an analysis is able to solve the exercise, judgement becomes clouded. The numerical value found for the Archimedean thrust via the standard method is scarcely affected by the offending assumption, and this example is a textbook case. If the mean internal and external pressures are stated as being equal, the error in their values is actually very small, and barely alters the result of calculating the corresponding densities; on the other hand, they depend strongly on temperature. Dare one say it, but the Archimedean thrust does rather well. However, the difference in pressures is conceptually essential. Although certainly very small, this difference, multiplied by hundreds of square meters of the envelope, does explain the lift.

The value of comparing several approaches to a given phenomenon is not, therefore, an hypothesis without foundation. Note, once again, that such a process requires an effort.

## 6.4 Testimonies of intellectual satisfaction

In the face of increasing demand, it becomes necessary to assess what the benefits are of going slightly beyond the conventional teaching rituals. Here are a few elements for assessing the effects of individual (interviews) or collective sessions (within the standard university context), limited in time to half an hour, and devoted to

74 VIENNOT L. (2006) Teaching rituals and students' intellectual satisfaction, *Phys. Educ.* **41**, 400-408 (http://stacks.iop.org/0031-9120/41/400); MATHÉ S. & VIENNOT L. (2009) Stressing the coherence of physics: Students journalists' and science mediators' reactions, *Problems of education in the 21st century*, **11**, 104-128; and Appendix E.

75 We limit ourselves to GIANCOLI's citation, given in note 70.

76 With just one exception: 129 out of 130.

questioning and discussing the previous ideas of this chapter.[77] The academic levels of the individuals concerned ranged from the first year at university to the second year of IUFM (*Instituts Universitaires de Formation des Maîtres*, Teachers' Training College) for trainee level 2 teachers (French PLC2). It is important to point out that the section to which the students belonged was not at all selective, and that they were not even all destined to continue in physics.[78] As has already been said, practically none of these students spontaneously spotted the absurdity of the offending assumption. However, at the end of a discussion on this subject a significant set of reactions were voiced:

During a series of 15 individual half-hour interviews in the first year of university, *all* the students involved reckoned the discussion was accessible, and that it was important to have had it despite the time it took. Their reactions went from satisfaction to looking back in annoyance.

Satisfaction was expressed simply and strongly:

- Thanks very much, I've learnt... a lot!
- That didn't seem to take long.
- Thanks. If you have other handy ideas like that, I'll be back!

This satisfaction should not convey the illusion of their having understood everything, nor that everything could in turn be immediately explained to a fellow student. It seems to be linked with the feeling of having a worthwhile and stimulating thought-provoking objective, and the feeling of time well spent:

- You realise that it's much more interesting to have understood... Developing a critical faculty is more important in my life.
- OK, explanations, you can't just give them like that, you made me think even though it was hard work; thinking is good for you, you learn a lot.
- You made me think, thanks!

Annoyance comes from having wasted time in the past:

- Ah yes, it's really worth it, that's what's interesting, otherwise, well, we're no more than robots.
- For example, I'm in the first year, one of these days I'm going to do some research, then I come back to it, I find an assumption in the exercise which is wrong, so does that mean that all I did in the past is based on erroneous assumptions?

77 Viennot L. (2006) Teaching rituals and students' intellectual satisfaction, *Phys. Educ.* **41**, 400-408 (http://stacks.iop.org/0031-9120/41/400); Viennot L. (2006) Modélisation dimensionnellement réductrice et traitement "particulaire" dans l'enseignement de la physique, *Didaskalia*, **28**, 9-32.

78 We observe similar reactions in a group of third year university students training to be journalists or scientific writers: Mathé S. & Viennot L. (2009) Stressing the coherence of physics: Students journalists and science mediators reactions, *Problems of education in the 21st century*, **11** (11) 104-128 and Appendix E.

- Of course I prefer that, what's the point of an exercise if you don't understand it!
- How come this is the first time anyone has told me that?

Note also the realistic attitudes of these students who provide feedback, either positive or negative for their teachers' role:

- (*More interesting?*) Absolutely, from the moment we are taught to do it.
- Personally, if I have an exercise to do, if there's an assumption in it, not in an exam, but afterwards, if I have the time, I'll do it and think about it.

A degree-level session (university third year) produced the same sort of assessment–yes, it's worth spending the time (the same as in the interviews, *i.e.* roughly half an hour)–among 18 of the 21 students present, of whom 17 added (when asked explicitly) that they had really enjoyed it (marked 3 or 4 on a scale of 1 to 4).

Note that it is not just a matter of enthusiasm sustained by the novelty of the subject, "dazzling" applications or "incredible" revelations. If it cannot be excluded that personal interaction may be a favourable factor during the interview, similar reactions were observed in group sessions as well.

Dare we suggest that this pleasure derived from reasoning, from the impression that just some of the rational elements of a judgement of formal consistency had been mastered?

## 6.5 Yet more links? Weight and pressure of the gas

Either on the level of questioning, or the experience of the deep pleasure of in-depth understanding, we can take up the challenge. Once started, the questioner quickly finds the tables are turned and he is required to go much further. Strictly speaking, one might object that the dependence on altitude of gas pressure is not linear, but is a decreasing exponential; this is used by a classic exercise at the start of university as a nice introduction to differential and integral procedures, and furthermore, the assumption that the gas in a balloon is at equilibrium, between two blasts from the burners, is somewhat audacious. Of course. So why, what's the point of making up such stories? In terms of the assumption discussed above, the answer is that it's not a question of some obscure fourth decimal place to be taken into account (or not) for more (or less) accuracy, but rather a question of recognising or denying the whole underlying principle of the phenomenon.

We now pursue an equally fundamental question: what is the weight of a gas?

It is often asserted without prior discussion that a column of atmospheric air exerts on the ground a force equal to its weight. Even if we omit to say so, this is an equilibrium situation. The argument is simple: a Newtonian force balance on the "column of air" system, followed by application of NEWTON's third law. However, this simplicity and the unquestionable Newtonian argument fail to resolve a question: O.K., we all agree, "it should...", but how does this actually *happen*? How do the molecules striking the ground "know" the weight of all the other molecules above them? How is it that the pressure due to the impacts of the molecules on the ground corresponds exactly to that dictated by a Newtonian force balance, *i.e.* the value obtained by dividing the weight of a column of air by the area of its base? This, if we think about it, is as surprising as it is routinely applied. Appendix B provides a suggestion for making this result more comprehensible mechanically. Echoes that we got back, particularly from trainee teachers, were sometimes disturbing. Phrases such as: do the molecules exert the same force on the ground as if all those in the column were stationary and touching? Some results from the survey, reproduced in Appendix B, lead us to suspect to what extent the answer (yes) is counter-intuitive. In the suggestion provided to clarify the issue we see the factor $g$ (gravitational acceleration) slipping in, which is never introduced in the usual courses on gases; this discretion is rarely justified[79]. Inevitably, stirring up this sort of questioning can have a domino effect: "But then if atmospheric pressure varies, as the national weather service tells us every day, is the result still true?" The vision of the atmosphere in equilibrium then has to be reconsidered.

If we push the need for reasoning a little, important points often resurface, tying up with what has already been discussed.

## 6.6 The advantages of changing the scale of analysis

This example, which sets a global analysis against another more local analysis (particle-based, even), by no means exhausts the subject it illustrates any more than the others do. In a wider sense, this ability to move from one scale to another in describing phenomena is of often crucial benefit in terms of understanding. In particular, the advantage of an intermediate scale between the macroscopic and the nanoscopic scales of atomic and molecular phenomena is now well established. This "mesoscopic" scale allows local values of variables to be defined, pressure and temperature for example, while averaging out molecular effects otherwise impossible to deal with. In hydrodynamics for example, it seems we have no choice since the unit of reasoning (if we can call it that) is mesoscopic: this is the "element of fluid volume", but it is less common to use this scale in hydrostatics. However, this allows, for example, a better understanding of the fact that the pressure on the ceiling of a submerged cave

79 This is not even the case for a chapter entitled "The air around us" in the French *Seconde générale* programme (Year 11) (2000: © MENRT, CNDP and GTP of physics-chemistry, 1999).

(Fig. 6.2) is non zero; as is shown from a detailed analysis of the repulsive interactions between the fluid or solid elements of the system.[80]

It is not common to treat friction in such a way as to placate certain intellectual dissent: "The ground is horizontal, smooth, and cannot provide motive force (for a walker)". Though this may be done by invoking mesoscopic asperities between ground and sole of the shoe, a view which recalls the starting-block situation, reconciling intuition and formalism.[81]

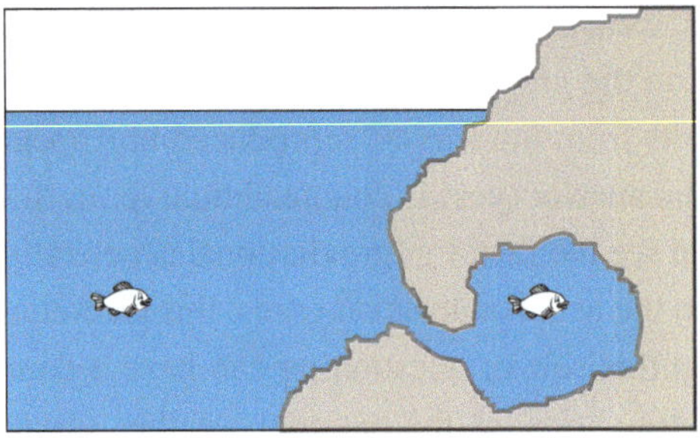

***Figure 6.2*** *- The two fish situation for a taxing question: is the water pressure the same for the two fish?*

We can say that each time a point in the cave "should be..."[82] at the same pressure as the point in the open sea at the same depth, or that the ground pushes the walker forwards, as per the Newtonian force balance. However, the reconciliation between these necessities of a macroscopic analysis and a more mechanistic view at the mesoscopic scale is of great benefit, if feedback from pupils is to be believed.

Going through the range of scales in the other direction, the macroscopic illustration of the effects involved at a much reduced scale is, at the very least, a powerful stimulus to thought. Still on the subject of friction phenomena, the teaching team at

**80** Situation suggested in PUGLIESE-JONA S. (1984) *Fisica e Laboratorio*, vol. 1, Loescher, Turin, and made use of in his thesis by BESSON U. (2001); see also several publications in the note which follows). It is often suggested that the rock does not push on the water, and in particular that there is no "water above" the level of the cave's roof.

**81** VIENNOT L. (2003) *Teaching Physics*, Dordrecht: Kluwer Academic Publishers, Chapter 3. On BESSON's study, see Chapter 3 in the same work. On a plea in favour of a mesoscopic approach, see also: BESSON U. & VIENNOT L. (2004) Using models at mesoscopic scale in teaching physics: two experimental interventions on solid friction and fluid statics, *International Journal of Science Education*, **26** (9) 1083-1110.

**82** Ugo BESSON underlined this idea in particular, see refs. in the previous note. In electromagnetism, several authors have been keen to establish a link between causal intuition and the "it should be..." of stationary states; see especially: CHABAY R.W. & SHERWOOD B. (2002) *Matter & Interactions II: Electric & Magnetic Interactions*, New York: John Wiley & Sons; and PSILLOS D. (1995) Adapting Instruction to Students' Reasoning. In D. PSILLOS (Ed.) "*European Research in Science Education*". Proceedings of the second PhD Summerschool. Leptokaria, Thessaloniki: Art of Text, 57-71. See also: CLEMENT J. (1993) Using bridging analogies and anchoring intuitions to deal with students' preconceptions in physics, *Journal of Research in Science Teaching*, **30**, 1241-1257.

Pavia therefore suggested, an introductory experiment on the dissipation of energy.[83] In normal conditions, a trolley launched against a wall rebounds in a manner resembling a quasi-elastic collision. This however, is no longer true if oscillating blades are mounted on the trolley; in this case the collision results in a quivering or a vibrational movement of the blades while the trolley supporting them no longer rebounds off the obstacle. Much remains to be explained about the dissipation of energy, an internal phenomenon of the body (or bodies) which heats up, or otherwise accompanied by the transfer of heat between the bodies and the outside. However, this experiment provides an excellent point of departure for tackling this somewhat mysterious topic.

In brief (and pupils confirm this), there is often a great deal to be gained by presenting several different points of view, or scales of analysis, in order to cast light on the same phenomenon. Let us hope that the examples discussed here have pleaded effectively in favour of this idea. Be that as it may, its importance is far reaching, and teachers are free to choose how best to illustrate it.

83 Besson U., Borghi L., De Ambrosis A. & Mascheretti P. (2007) How to teach friction: Experiments and models, *American Journal of Physics,* **75** (12) 1106-1113. Video available at http://fisica.unipv.it/didattica/Energia/ENG/irrevers.htm

# Part III

## Simplicity:
## Ruin or triumph of coherence?

# Chapter 7

## OPTIMISING SIMPLE EXPERIMENTS

### 7.1 Can we rely on simplicity?

The question is at least worth asking: in relying on so-called evidence or an obligation to simplify, is there not a danger of losing the flavour, or even the very substance of physics? Illustrated above by apparently simple questions, this question can be reversed by a stimulating suggestion: with regard to the most insignificant question, a consistent line of physical reasoning may offer the support for a conceptual advance and, dare we say it, intellectual satisfaction. Depending on the manner in which it is set out, "simplicity" may be revealed as being either suspicious or, quite the contrary, very useful.[84]

In what follows these two faces of "simplicity" are again illustrated by attractive "mini-demos", since these days attractiveness is the key word. To spare the reader too much in the way of theoretical meanderings, this illustration will draw on situations related to elementary physics that have been already discussed: the mechanics and statics of fluids.

The pleasure engendered by many of the "mini-demos" depends on their surprising result, "counter-intuitive" some might say.[85] At the first level of analysis, hopefully,

---

84 Concerning "suspicious simplicity", the book by D. KAHNEMAN (2012) *Thinking Fast and Slow* (London: Penguin books) is very instructive. Concerning virtuous simplicity in physics, see note 10 and also this excerpt from OGBORN J. (2010) Science and Commonsense. In M. VICENTINI and E. SASSI (Eds.): *Physics Education: recent developments in the interaction between research and teaching*, New Delhi: Angus and Grapher Publishers (http://web.phys.ksu.edu/icpe/Publications/index.html): "If we want to try to see some imagined entity acting alone, we have to limit and control the actions of entities which may disturb or conceal the behaviour of the first. Thus we are driven, in BACON's words, to vex Nature, deforming natural states of affairs so as to simplify them and to disclose more clearly the behaviour of a given entity."

85 For a very long time, and far from being content with the effect of surprise, some highly studied versions of such experimental proposals included a real content. Hence: MCDERMOTT L.C. *et al.* (1996) *Physics by Inquiry*, vol. I and II, New York: John WILEY & Sons; CHAUVET F. (1996) Teaching colour: designing and evaluation of a sequence, *European Journal of Teacher Education*, vol. **19**, n°2, 119-134 and VIENNOT L. (2003) *Teaching physics*, Dordrecht: Kluwer Ac. Pub., Chap. 6.

the surprise might generate interest and at the same time encourage young students to embark on a journey into science. More simply, it might improve the image of science and scientists. To say that the result of an experiment is astonishing is the same as saying that the prediction which springs to the minds of non-specialists is wrong. Before pursuing this line, it could be useful to reflect a moment on the analysis of such "errors".

## 7.2 ARCHIMEDES' scales

The experimental setup shown in Figure 7.1(a) contains an implicit question: if a ball of modelling clay is immersed in the water, held by a thread half-way up the container, will the platen on the scales supporting it go down, up, or stay at the same level? The proportion (the majority) making the incorrect prediction, that the platen stays at the same height, is commonly around 80%, even among physics "instructors".[86] Actually, the platen goes down significantly. The submerged ball experiences a so-called "Archimedean" force due to the water, and the water-container ensemble therefore experiences, by virtue of NEWTON's third law, an opposing force due to the ball. Before the ball is immersed, at equilibrium the water-container ensemble undergoes its proper weight and the opposite force exerted by the platen. The scales show the weight of the container full of water. After the ball is immersed, the water-container ensemble, experiencing an additional downwards force, cannot reach equilibrium without an increase in the upwards force which the platen exerts on the container. The motion of the platen reflects the unbalance thus created, generating the increase in the force required for equilibrium (Fig. 7.1(b)).

When this equilibrium is reached again, the force interacting between the platen and the water-container ensemble is greater than that of the weight of the whole. The reading on the scales no longer corresponds to the weight of the water-filled container, but a greater value.

If we are interested in finding the roots of this common error–beyond the simple statement that "it's counter-intuitive"–then, based on previous research, several non-exclusive hypotheses arise.

---

Recent movements falling within the "Inquiry Based Science Education" all claim to have this same ambition. Hence ALLENDE J.E. (2008) Academies Active in Education, *Science*, 321, 29-8-2008. Editorial: "(…) science education that is based on inquiry, an approach that reproduces in the classroom the learning process of scientists: formulating questions, doing experiments, collecting and comparing data, reaching conclusions, and extrapolating these findings to more general situations. See also: ROCARD Y. (2007) *Science Education Now*, Report EU 22-845, European Commission, Brussels (http://ec.europa.eu/research/science-society/document_library/pdf_06/report-rocard-on-science-education_en.pdf).

86 Doctoral students training to be teachers, University Paris Diderot, numbers amassed over 15 years (1994-2011): 340.

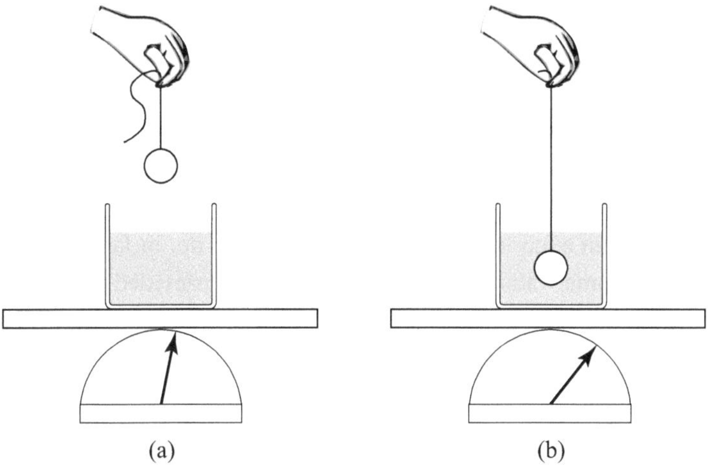


***Figure 7.1*** *- The* ARCHIMEDES' *scales experiment*

(a) *The reading on the scale indicates the weight of the [water+container] system.* (b) *Once the ball is immersed, the reading on the scale is greater than the weight of the [water+container] system*

For instance, it could be suggested that everybody believes that "an object (always) exerts its weight on its support", a pseudo-rule refuted by the water-filled glass (Fig. 7.1(b)).[87] The root of the error would then in this case be broader than simple ignorance.

And, one might add that the action-reaction law has been pretty badly mistreated, since ignoring the action of the ball on the water is tantamount to contradicting this law. The error in predicting the motion of the platen therefore joins numerous other cases.[88]

A more psychological formulation might tie up the fact of neglecting NEWTON's third law with a mental Agent/Patient[89] type of scheme, introducing a constitutive asymmetry into the interaction.

87 Note that this counter-example involves the static situation. In the event of the support accelerating (see the lift problem) the students have less problem in seeing that the object does not always exert its weight on its support (this approximate formulation paraphrases the common usage).

88 VIENNOT L. (1982) L'action et la réaction sont-elles bien égales et opposées ? *Bulletin de l'Union des Physiciens*, **640**, 479-485; see also: VIENNOT L. (2004) *The design of teaching sequences in physics. Can research inform practice? A Lines of attention. Optics and solid friction.* In E.F. REDISH & M. VICENTINI (Eds.) Research on Physics Education. Course CLVI, SIF Varenna. Amsterdam: IOS press, 511-513. BROWN D.E. (1989) Students' concept of force: the importance of understanding Newton's law, *Physics Education*, **24** (6), 353-357.

89 ANDERSSON B. (1989) The experiential Gelstalt of Causation: a common core to pupils preconceptions in science, *European Journal of Science Education*, **8** (2) 155-171.

Like concentric circles, these formulations progressively widen the candidate interpretations. We will come across these further on. In the second example we would like to go on and highlight how they can actually be of service.

## 7.3 The inverted glass of water

Figure 7.2 shows an experiment which is very easy to do, in fact nothing could be simpler: a glass, some water, and a piece of card. And, provided that the card is sufficiently wet, the water doesn't escape when the whole thing is turned upside down.

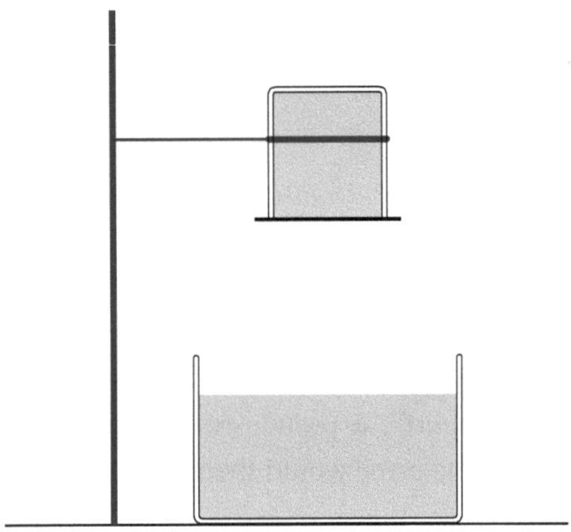

***Figure 7.2*** *- A glass of water covered with a card and turned upside down*

In reaction to this, the following explanation is not uncommon: the card does not fall because the atmosphere "supports the water's weight". This explanation makes use of two relevant forces, but it suggests a Newtonian balance between them, whereas, in fact the upward force on the card is about a hundred times greater than the weight of the water. That is to say, the above explanation is, at best, very incomplete, and at worst, quite misleading.

The atmosphere pushes the card upwards, thereby preventing the water from falling. And quite simply it can even be added that the action of the atmosphere on the card is one hundred times greater than the weight of the water.[90] Does this mean that we can conclude that the upward force on the card is sufficient to support this load?

**90** International teacher training workshop, May 2006, private communication.

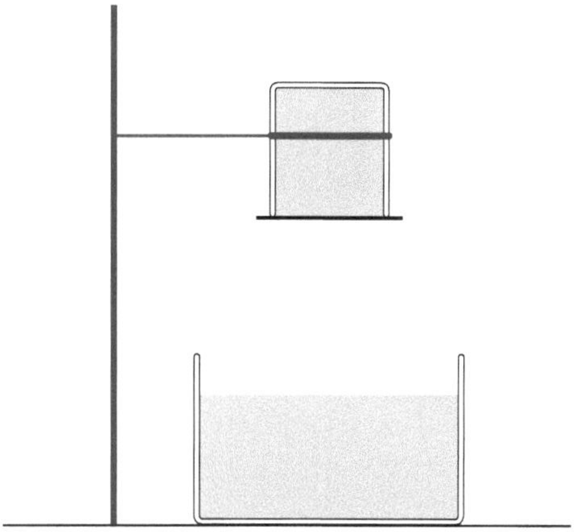

***Statements often found in typical, common place explanations:***[91]
The water exerts on the card a force equal to its weight. The force due to atmospheric pressure supports the card which therefore does not fall down.

***Opposite, to the right:***
Diagram (not to scale) illustrating the order of magnitude of the forces mentioned in the typical, commonplace explanation.

***Figure** 7.3 - A simple experiment that often gives rise to a problematic explanation*[92]

This case is different from the previous one. Apart from a potentially incorrect prediction (the water will fall), the typical, commonplace explanatory comments merit analysis.

The card experiences the upwards action of the atmosphere, and only experiences a downwards force equal to the weight of the water; this is a hundred times less, so no equilibrium can be possible. To suggest that there might be violates consistency if NEWTON's second law is to be believed.

It is therefore false to assert that the card is subject to the downward force of the "weight of the water". It experiences a much larger force, opposed to and of the same order as that exerted by the atmosphere (Fig. 7.4).

To understand the genesis of the typical, commonplace explanation, we may advance the existence of now familiar possible sources:

- It is widely believed that "an object (always) exerts its weight on its support", a pseudo-rule here negated by the upturned glass.
- The action-reaction law is ignored. To assume that the card experiences a large force due to the atmosphere inevitably raises the question of the reaction; this would be impossible to ensure if this light object itself experienced only a small downwards force due to the water.
- An interpretation of the situation in terms of Agent (atmosphere) and Patient (that which risks falling) might account for the drastically reduced explanation analysed here.

91 For example these elements are in the write-up for the workshop cited in the previous note.

92 VIENNOT L. (2010) Physics education research and inquiry-based teaching: a question of didactical consistency. In K. KORTLAND (Ed.). *Designing Theory-Based Teaching-Learning Sequences for Science Education,* Cdβ press, Utrecht, 39-56 and VIENNOT L, PLANINŠIČ G., SASSI E. & UCKE C. (2010) *Various experiments involving fluid statics* (www.eps.org, select *Education* and then select *MUSE*).

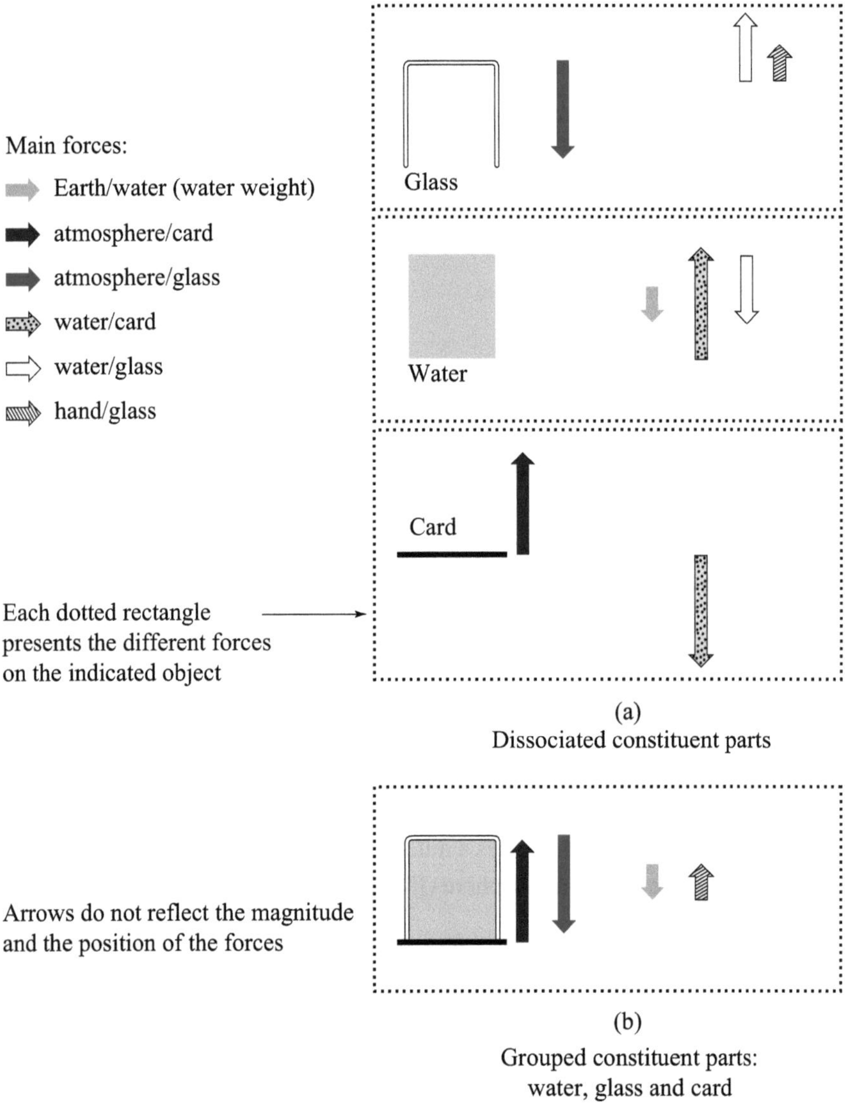


***Figure 7.4*** *- Main forces (vertical components) in the situation of the glass full of water held upside down (for more details, see* WELTIN *1961,* VIENNOT *et al. 2009), are shown by considering its constituent parts separately (a) or grouped together (b). To simplify the diagram, the interaction between glass and card,*[93] *the weight of the card and that of the glass are not shown. These forces, even though the primary one may play an important role, are small in comparison with the others. Including them does not endorse the idea that the card or the atmosphere are doing nothing except "supporting" the water.*

**93** This can be thought of at first as repulsive (solid-solid) or attractive (via capillarity). Various filling situations with some air within the glass can lead to an analogous equilibrium, in this connection see WELTIN H. (1961) A paradox, *Americal Journal of Physics*, **29** (10) 711-712.

We may add to this list another seemingly trivial common characteristic of the explanation but one which is probably the most important and the most widespread.

- To analyse an interacting system of objects, we start from an end where something is "going on"[94] and we often go no farther.

It is at the lower part of the glass where there is most chance of something happening. If we look only at this, the essential feature escapes analysis. Figure 7.4 (from the point of view of their values)[95] sums up the main forces present. It appears that the weight of the glass-water ensemble and the opposing action of the hand holding the glass are of little importance compared with the compressive interactions (air-card, card-water, water-bottom of the glass, bottom of the glass-air) related to the atmosphere: It is here that the central phenomenon lies.

For this example, as for the previous one, the reasons which may explain the errors or typical, commonplace explanations are the same. The point raised here is an idea which will have an important place in the last part of this book: when explaining certain phenomena even physicists frequently draw upon lines of reasoning which echo those of non specialists. The coherence of the discipline does not emerge strengthened from this game of mirrors.

## 7.4 The inverted test-tube

A test-tube, containing water, turned upside down over a bowl full of water, is essentially no different from the analogously demonstrated glass. At the end (the bottom) of the test-tube is the free surface of the bowl at atmospheric pressure. Water is contained there in the test-tube. Above, with or without an intermediate layer of gas, is the end of the test-tube held either by a hand or some other support. The forces at work shift from one situation to the other, and only their relative magnitudes can alter. The higher the column of water, the less is its interaction with the upper end of the test-tube. We know that for a ten metre column of water this interaction falls to practically zero, and if the glass tube is raised still further, the vapour pressure of water at that temperature determines and limits the very weak interaction between the end and the contents of the test-tube.

The equivalent of the misleading idea ascribing the support of the water in the inverted glass to the atmosphere is, in this case, that it is the force due to atmospheric pressure at the surface of the bowl which balances out the weight of the column of water in the test-tube. Other forces are to be taken into account. For example, with two metres

94 In 1981, S. Fauconnet emphasised the importance of this phenomenon in his thesis: *Etude de résolution de problème : quelques problèmes de même structure en physique* ; LDSP, Université Paris 7.

95 Capillary forces in particular have an important role to play in preventing the outside air from getting in past the card, but their magnitude is small in relation to the other forces involved.

of water above the level of the bowl, the compression at the top of the test-tube, at equilibrium, would correspond to four fifths of atmospheric pressure (Fig. 7.5).

Read what Isabelle CHAVANNES has to say about Marie CURIE's lessons to her friends'children on this situation:[96]

"It (the water) remains at the top of the tube. What sustains the water in the tube, what holds up this 2 meter column of water? It's the atmospheric pressure pushing on the water in the container. There is no air in the tube, and no pressure is exerted on the water."

We find similar sorts of explanations as those given for the upturned glass. The idea of support is entailed in the term "sustaining". What is responsible for this support, or holding up, is the atmosphere, which implicitly compensates for the weight of the water. Things are happening at the bottom of the column, while the interaction between it and the glass at the top of the tube is, in complete disregard of the requirements for Newtonian equilibrium, blatantly ignored.

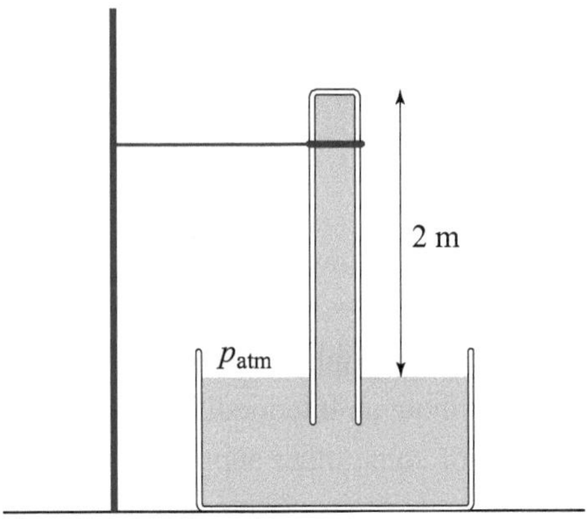


***Explanation given:***

"What holds up this 2 metre high column of water? It's the atmospheric pressure that is pushing on the water in the container. In the tube, there is no air, and no pressure is exerted on the water." (reference in note 95)

***Opposite, to the right:*** Diagram (not to scale) illustrating the order of magnitude of the forces acting on the column of water, as mentioned in the attached explanation.

***Figure 7.5*** *- A similar situation to that of the water-filled inverted glass: a test-tube over a bowl of water. The explanation given (see text) has the same features as the customary explanation for the glass.*

This is a NOBEL laureate talking, albeit through her student. That's to say that in terms of explanation the limits already emphasised are not random or marginal phenomena, nor details falling under sterile purism. These limitations undermine the very consistency of the idea.

This is a topic which we will come back to later. The phenomenon is too important and too much neglected to have qualms about overstating it. However, the aim is not just to take pleasure in denouncing it. At little cost, we can reorient the design of an experiment which might otherwise be exposed to criticism while, at the same time, very probably altering its effect on the participants.

96 CHAVANNES I. (2003) *1907. Leçons de Marie CURIE aux enfants de nos amis,* EDP Sciences, Paris, p. 46.

## 7.5 Beyond rituals

We have already pointed out that the Archimedean submerged ball experiment as presented above runs counter to customary teaching practice. Stress is frequently laid on the fact that, once immersed, the force exerted by the ball on the thread supporting it is lessened. The lessons recorded by I. CHAVANNES frequently mention the "loss of weight" of an immersed body.[97] And in fact, we practically always talk of the Archimedean thrust, with that of the water acting as "agent" on the ball ("patient"), which "experiences" or even "receives" it.[98] On the contrary, we may emphasise that there is an interaction, as in the 1992 French *troisième* (Year 10) programme. The glass on the scales situation (Fig. 7.1) highlights this much neglected opposing force. Hence we can learn something new from this, in particular if the expression "Archimedean interaction" is being "bandied about".

As for the glass of water experiment, a minor modification gets rid of the above mentioned dangerous ambiguities. In this question, which is in fact dominated by atmospheric compression, the problem of the support could potentially constitute a major interference. Consequently, we will place the glass horizontally (Fig. 7.6) and analyse the components of the forces in this direction. In the horizontal position, the atmosphere plays the same role as when vertical, *i.e.* the main role, compression. Abandoning the ritual vertical arrangement enables us to focus on the essential features[99] and this is neither complicated nor costly.

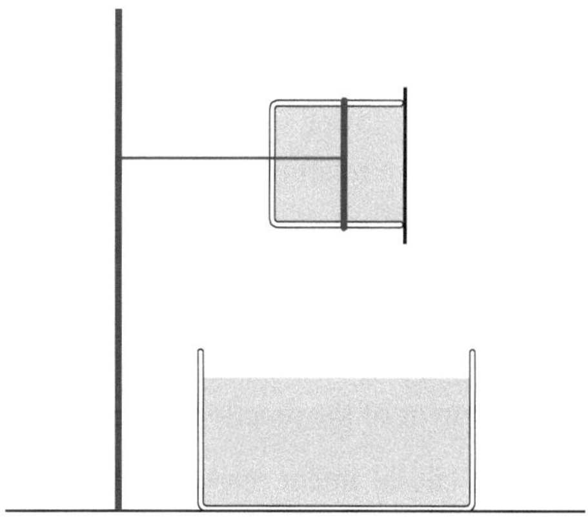

***Figure 7.6*** *- Placed horizontally, the water is contained just as well as when vertical, as in Figures 7.2 and 7.3.*

97 For example p. 64: "(...) in the water it loses part of its weight".

98 For example: ALLÈGRE C. (2006) *Un peu plus de science pour tout le monde*, Odile Jacob, Paris, p. 31.

99 The result of this modified experiment can surprise even a physicist (as noted by the author, but not formalised by research).

## 7.6 Echo-explanations and causal linear reasoning

In so far as prediction or explanation are concerned, the two previous examples illustrate the de facto complicity between the typical, commonplace mind sets of supposed experts and those of novices. It is at this point that we would like to introduce a useful term, namely "echo-explanation", to remind us of this proximity. In so doing, we did not intend to take position with regard to the intentions of the "supposed expert", who is defined as anyone who might be thinking of giving an expert explanation. Thus, it is quite possible, for example, that Marie CURIE chose to talk only about one end of the column of water suspended over the bowl, even at the expense of seriously stretching Newtonian consistency, so as not to "complicate" the issue. What we are trying to do here is to emphasise just how frequently explanations are passed off as expert, when they have exactly the same substance as those of non-specialists. These "echo-explanations" carry the same limitations as those which mark typical, commonplace ways of reasoning.

The most meaningful explanatory structure in this regard is without a doubt that of "causal linear reasoning". Appendix C reviews the foundations of this model of reasoning, and illustrates the contrast between it and the model of "quasi-static" or "quasi-stationary" changes so widespread in physics. It is thanks to this contrast that causal linear reasoning is of such benefit in interpreting the learning difficulties of students.

Imagine, therefore, a system consisting of two springs (of known unstretched length and stiffness) end to end suspended from the ceiling; the bottom end is gently pulled. The system can be described by several variables (spring length and tension, total length, external force on the lower spring). These variables are connected by simple relations, for instance those relating the total length to those of each spring. When the system stretches as a result of an externally applied tension, the so-called "quasi-static" analysis of the system consists of assuming that the values of the various variables all change at the same time, while continuously satisfying some simple relationships. These enable a certain number of questions to be resolved, for example, that of the displacement of the junction point for a given total extension.

In a case like this, a typical causal linear reasoning involves focusing initially on what is happening at the lower end of the system. Associating, for example, the external force, the stiffness of the lower spring and the displacement of the lower point, which in its turn is all too frequently associated with the extension of the lower spring (which is less). "The first spring will extend. Then, after a while, the second will also extend", as one pupil wrote.[100] Even without the visual help of a line or a

100 Symptomatic of causal linear reasoning, and contrary to the trail of calculations for this view of the situation, such an explicit statement is not often encountered. See FAUCONNET S. (1981) *Etude de résolution de problèmes : quelques problèmes de même structure en physique*, Thèse de troisième cycle, Université Paris 7, p. 112.

circuit, which suggests a quasi-geometric route for thought, we frequently observe explanations which start off from a simple phenomenon or from the change in a single variable, and then carry on with a sequence of "cause and effect" determinations.

This is not a question of variables changing all at the same time under the permanent constraint of simple relationships (e.g. $pV = nRT$ for ideal gases)[101]but of a sequence of changes involving one variable at a time, each one in turn. A typical linguistic connector for what, to all intents and purposes is a story, is the word "then": "The volume gets smaller, then the density increases, then the number of impacts goes up, then the pressure goes up",[102] or, "You heat the gas, so the temperature goes up, then the pressure goes up, then the volume goes up".[103] Independently of the associated risk of error, we can observe the narrative structure of the typical, commonplace explanatory comment.

Over and above the specific errors which are caused, when causal linear reasoning fails this is always because several relevant aspects of the system are not accounted for at the *same time*. Even when restricted to the simple case of quasi-static transformations,[104] it is not possible to determine what the gas pressure becomes without at the same time considering its volume or temperature, or without *simultaneously* considering the transfer of heat and mechanical energy involved. Taking the simplest situation of the two end-to-end springs changing slowly, we cannot know what happens to the lower spring when pulled downwards without knowing the properties of the upper spring.

After this brief review, let's go back and pick up the thread of our simple experiments. The reader will have noted in passing that the focus on a single end of the system has already held our attention. The inverted glass of water, the test-tube holding a column of water above a bowl: two opportunities to observe that the analysis is commonly restricted to an opposing force at the bottom, where there is a danger that the water will escape.

Now let's take the example of a more dynamic event, definitely better for the narrative: siphoning water from a bowl.

101 See above: Chapter 3, note 46.

102 Commonly given for a quasi-static adiabatic compression, this explanation says nothing about temperature. Its conclusion could be invalidated experimentally by simultaneously putting the system in contact with a cold source. See Rozier S., Viennot L. (1991) Students' reasoning in thermodynamics, *International Journal of Science Education*, **13**, n°2, 159-170; Viennot L. (2001) *Reasoning in Physics - The part of common sense*, Dordrecht: Kluwer Ac. Pub., Chapter 5.

103 Explanation commonly given for a quasi-static isobaric transformation. See references in previous note.

104 This adjective signifies precisely that each state of the system may be likened to a state of equilibrium, allowing the "continuous" use of simple relationships relating to this case (hence the ideal gas relation).

Marie CURIE gives us her version, via her pupil Isabelle CHAVANNES[105] : *The water contained in the long arm of the siphon flows out. A vacuum is created and atmospheric pressure causes the water to rise in the short arm, which is immersed in the container.*

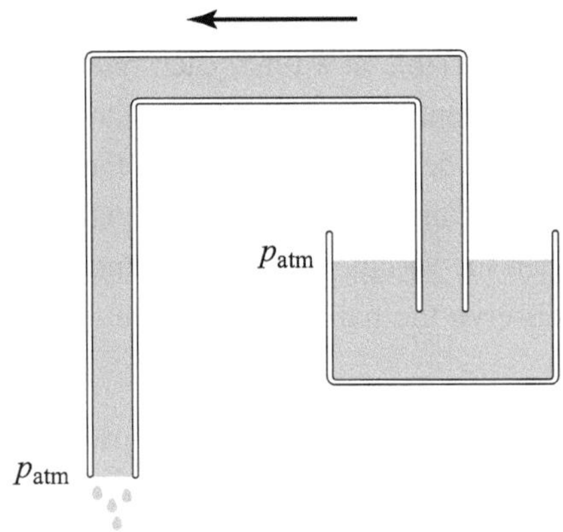


***Figure** 7.7 - A siphon situation*

The causal linear structure (now) becomes crystal clear:

*The water in the long arm of the siphon flows out → A vacuum is created → atmospheric pressure causes the water to rise in the short arm which is immersed in the container* (…).

To be sure, the fact of replacing an absence of connecting elements, or even the word "and", by an arrow is the fruit of a single interpretation. However, the most reliable criterion for declaring a structure to be causal linear is close at hand. Let's start at the end: if the atmospheric pressure causes the water in the bowl to rise up the small arm, this is because there is a vacuum (implicitly: above the bowl, somewhere in the tube). However, the end of the tube on the other side is also at atmospheric pressure. Well, what does this mean? We move on to the second link in the chain–A vacuum is created. This poses the specific question as to the whereabouts of this event and as to its somewhat temporary, not to say hypothetical, character.[106] The first link is an unjustified assertion that *the water contained in the long arm of the siphon "flows out"*. Is this to say that there is some proof, for example: without any support, does the water have to "fall out" (sounds familiar?)?

Note that this assertion is made independently of the whole system. We may talk about the *long* arm, but is it because it is *longer* than the other, or because it opens out into free air which is where there is a possibility of something happening?

105 CHAVANNES I. (2003) *1907. Leçons de Marie CURIE aux enfants de nos amis*, EDP Sciences, Paris, p. 62.

106 One frequently hears that if the water did not rise on the bowl side, it would create a vacuum, which is impossible.

If we do not realise that this *same* arm which opens out into the free air can give rise to a flow of water in the *opposite direction*, we have failed to understand anything about how a siphon works. We have only to put the bowl lower than the free air orifice (Fig. 7.8), and not a drop will escape on opening this orifice. On the contrary, the water will rise back up in this arm and will fill the bowl a little more.

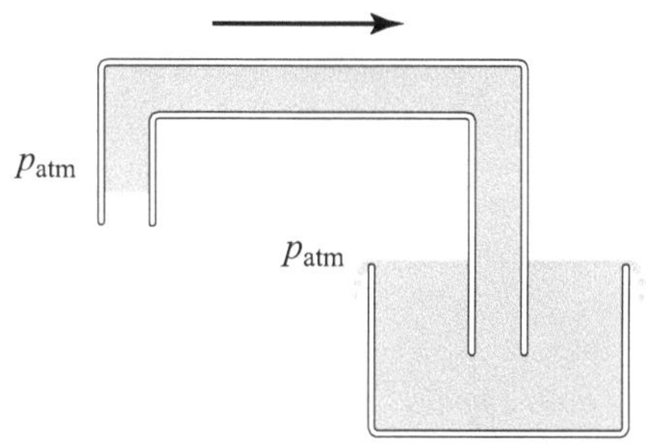


***Figure 7.8*** *- A siphon working backwards, shown just after releasing the orifice to the left of the figure.*

Typically, a siphon is a system in which one is obliged to observe both sides at the same time. To suggest otherwise is a blatant denial of the appropriate physical analysis: this is not just a minor detail.

However, common practice is wholly consistent with this reading of the phenomenon. See C. SANTAMARIA,[107] for example. On a perfectly ordinary drawing of a siphon arrangement, and one which emphasises the difference in heights of the free surfaces by the letter *H*, we read the following caption on the free air arm (in fact the longer arm): "The weight of the liquid on the right is sufficient to draw up the liquid in the left-hand tube", while to the side of the bowl (on the left on the drawing) this idea surfaces again: "the liquid is drawn up by the liquid on the right". Is "drawn" some magic word?

To be sure, in using the word "sufficient" the author has undoubtedly in mind a comparison[108] between what is on the side of the free air end and what is on the bowl side. However, if what concerns us is the reader of such a text, it has to be said that many links in the chain of reasoning remain unstated.

**107** SANTAMARIA C. (2007) *La physique tout simplement – Ne vous noyez pas dans un verre d'eau*, Ellipses, Paris, p. 28.

**108** Something else to consider with care: it might be sufficient that the tube is wider on the side of the bowl for a small weight of water on the side with the arm in the open air to "draw" (using the author's expression) a greater weight on the bowl side. In this regard see the instructive video suggested by G. PLANINŠIČ (http://www.fmf.uni-lj.si/~planinsic/PEMbG.htm).

*Static situation (axis pointing upwards)*

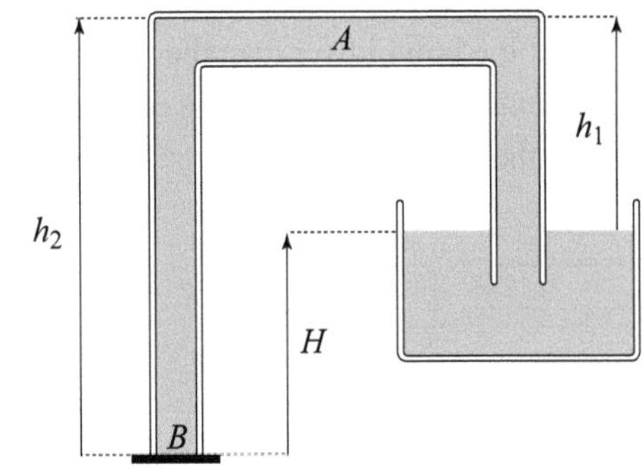


$p_A$, $p_B$: pressures at points $A$ and $B$ in the liquid

$$p_{\text{atm}} - p_A = \rho g h_1$$

$$p_B - p_A = \rho g h_2$$

herefore, since

$$h_2 - h_1 = H$$

$$p_B - p_{\text{atm}} = \rho g H$$

Here $H > 0$ and $p_B - p_{\text{atm}} > 0$

*After releasing end B, the bowl empties*

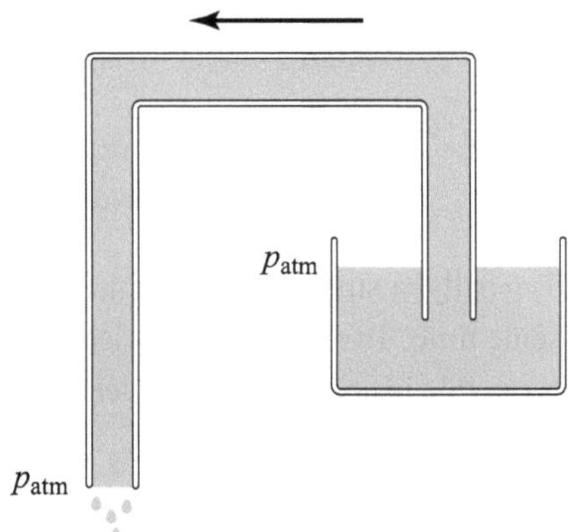


***Figure 7.9** - In the static situation, a pressure greater than atmospheric is exerted on the object occluding the lowest end of the tube, to the left in the figure.*

So, "mission: impossible"? Figures 7.9 and 7.10 give an analysis of the static situation before releasing the end of the tube, which will subsequently be in the free air. Simply writing down two hydrostatic balances governs the question. In the case of Figures 7.7 and 7.9 and because of what happens on the side of the bowl (free surface of the water higher than $B$), it turns out that the static pressure, before the end is opened to the free air, is greater than atmospheric pressure. Releasing this end abruptly lowers the pressure exerted on the water at that point, and the water column on that side is no longer in equilibrium since the pressure difference at its ends is too small. It thus falls. Simultaneously, the pressure at the top of the tube decreases, and the part on the bowl side (now not in equilibrium) moves upwards.

In the case of Figures 7.8 and 7.10, and on account of what is happening beside of the bowl (free surface of the water lower than $B$), the static pressure before releasing the tube is less than atmospheric pressure. Releasing this end abruptly increases the pressure at that point, and the water column on that side is no longer in equilibrium since the pressure difference at its ends is too great. This column of water rises.

Simultaneously, the pressure at the top of the tube rises, the column of water on the bowl side is not in equilibrium and moves downwards.

*Static situation (axis pointing upwards)*

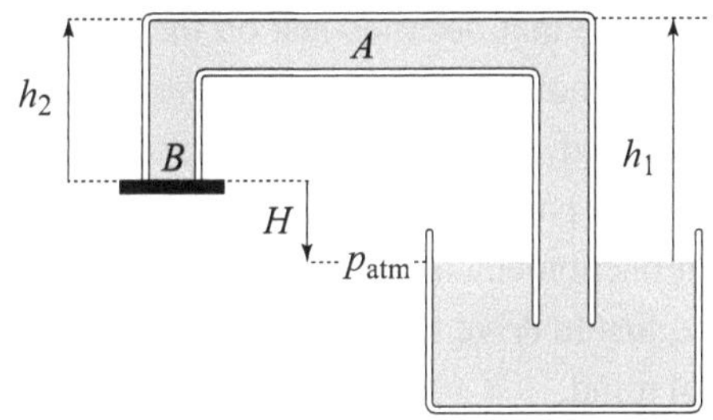


*Reminder:*

$$p_B - p_{atm} = \rho g H$$

Here $H < 0$

and $p_B - p_{atm} < 0$

*After releasing the end B, the water flows towards the bowl.*

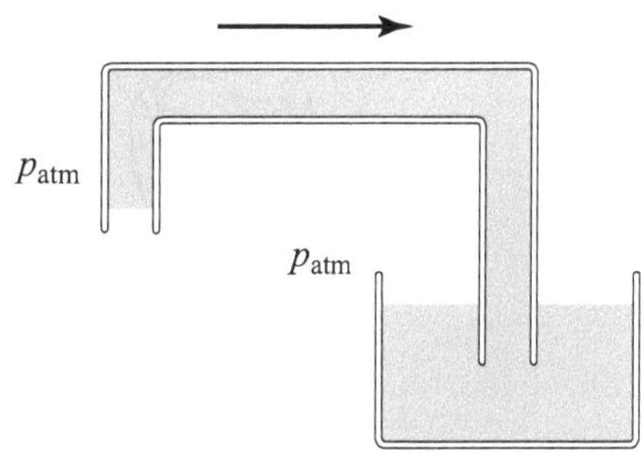


***Figure 7.10*** *- In the static situation, a pressure less than atmospheric is exerted on the object occluding the highest end of the tube, to the left in the figure.*

As for any departure from equilibrium, we can discuss the idea of simultaneity, which relates here to the timescale of our perception. However, consideration of only one side of the siphon is a far more serious distortion of standard physics.

## 7.7 When a simple experiment challenges simplistic reasoning

The glass of water, the test-tube and the siphon: these intentionally reduced watery experiments provide a framework for analysing what are in fact relatively general intellectual phenomena. They plead in favour of paying a great deal of attention to the aimed-for explanation, regardless of the apparent banality of the physical situation involved. We now return to how the experimental apparatus itself can be used to clarify those explanatory elements which we consider to be so crucial.

As we indicated earlier, the glass of water was set horizontally (as in Fig. 7.6). What can be done then for the siphon?

One suggestion consistent with the analysis developed in the previous lines is to present the arm of the apparatus, which is suspended in mid-air, full of water and blocked by a finger or some other plug; a screen conceals the rest, bowl and submerged arm. Question: what will the water do when the plug is removed, given that behind the screen no-one is blowing or sucking on the water, or even with a piston. We deceive expectation by raising or lowering the bowl behind the screen (Fig. 7.11). The only answer which can be considered right is the one that depends just on the (relative) height of the free surface of the bowl. The aim is not to lay a trap for its own sake, but so that the central idea can be properly introduced: that is to say nothing can be said about the behaviour of such a system if only one part is taken into consideration. No less important is the idea that, for physical phenomena anyway, it's the differences which "make the world go round".[109] Here, just to drive the point home, is one final example for an experiment with this idea in mind.

***Figure 7.11*** - *The screen hides the siphon.* A *can or not obstrue the siphon tip with her right hand, and move the water bowl up or down. If* B *says that the water will spill,* A *can prove him right or wrong!*

## 7.8 The "lovemeter"

A successful device, reflecting the poetic description of the title, is shown in Figure 7.12. Normal use is to increase the temperature of the lower reservoir to

109 Boohan R. & Ogborn J. (1997) designed a teaching programme highly relevant to this topic for secondary level pupils: Differences, energy and change: a simple approach through pictures, *New ways of teaching physics* - Proceedings of the Girep International Conference 1996 in Ljubljana, S. Oblack, M. Hribar, K. Luchner, M. Munih, Board of Education Slovenia. This idea is without doubt far too novel, and has never been implemented on a large scale.

generate an increase in pressure (the liquid's saturation vapour pressure), with spectacular effect.

This could be an exercise for the reader to guess the use into which it will be transformed. As we said above, and as the second law of thermodynamics tells us, it is differences that "make the world go round". A single source of heat never made an engine. In the first use envisaged here, we introduce a *difference* in temperatures, thereby increasing the initial pressure difference. We can do the same by projecting a cold jet of water on the upper part, and it's still an increase in the pressure *difference* which gets the liquid moving.

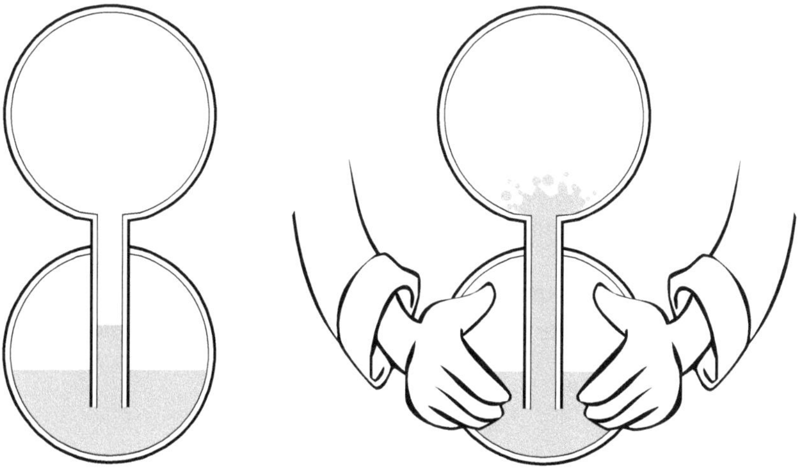

(a) (b)

***Figure 7.12*** - (a) *The gas is enclosed in each container between the glass wall and the surface of the liquid. In this equilibrium state, a slight pressure difference between the two containers maintains a small amount of the liquid in the central column linking them.* (b) *When the lower container is heated with the hands, the liquid surges upwards into the upper container; the same result is obtained when very cold water is poured onto the upper container.*

The expected conceptual benefit is what we were aiming for several pages back, *i.e.* to free ourselves from the minimal explanation which echoes the causal linear reasoning and to emphasise that it's the differences which matter here: one might add that as a bonus, it's a good way to explain the oscillating duck which never seems to get tired (Fig. 7.13).

***Figure 7.13*** *- Balanced on the edge of a cup full of water, the "duck" continues to oscillate until the water (and its beak) dries out.*

(a) *Because of a lower gas pressure in the upper container, it is partially filled with liquid; a large quantity of liquid in the beak makes the duck tilt over.*

(b) *The duck tips its beak into the water, allowing communication between the gases in the two containers via the central column; the gas pressures abruptly equalise, and the liquid flows back into the lower container.*

(c) *Evaporation of the water cools the beak resulting in both a temperature and a pressure difference between top and bottom which causes the liquid to rise again (and so back to case a).*

As we can see, the two experiments analysed above under the heading "lovemeter" are both useful in order to understand the operation of this mysterious toy (see Fig. 7.13).

## 7.9 Final remarks

Coming back to the title of this section of the book, these different experimental examples clearly bring out the gulf separating on the one hand, dearly paid-for simplicity and on the other, lines of approach which draw attention to the essential. For such a diversity of effects, what makes the difference is neither the experimental difficulty of the chosen scenario, not even the teaching time required. Putting a glass full of water in a horizontal rather than vertical position involves no extra cost. The problem lies in coming to acknowledge the need to reconsider these time-worn rituals and explanations of ours which reflect too much the short-comings of common sense.

What can help us attain our objective? It is by illustrating the high degree of consistency of physical theories in everyday occurrences and by helping students find pleasure in their work. If the aim is just to surprise, the previous thoughts might be considered little more than a waste of time. However, if the aim is consistency, or at least if we want to avoid compromising it, it is worth looking twice before presenting pseudo-explanations in which we pretend to believe. This, it should be pointed out, concerns not just those explanations which contain errors, such as the idea that the card which covers the inverted glass of water is subject to the downward "weight of the water".

In most cases, there is no clear-cut error in whatever is being proposed, only the persistence of misleading ideas, especially if the suggested explanations are reflections of common sense. Apart from habitual custom, this apparent off-handedness may well come from ignorance of the risks involved. The experiments involving the "material" light beams[110] pose problems mainly because we have a common tendency to imagine that light can be seen "in profile", like a passing train. The numerous occasions on which we think that an object "exerts its weight on its support" merit our attention and a targeted response precisely because so many people think that that is always the case. In positive terms this time, we can emphasise to what extent physical systems require more than one variable, more than one position or more than one local effect to be taken into account. So, cooling the top part of a "lovemeter" is not gratuitous fantasy, but goes to the very heart of the problem.

As teachers, among the things we may do inadvertently, the most counter-productive is without doubt that of providing an "explanation" which explains little or nothing. The siphons, where we started off by saying that the water flows out in the arm of the tube which is open to the free air, and which has an opening pointing downwards, are poorly "explained" if we don't know why this water should flow out. Where is the evidence? Would a pipe with a downwards pointing opening inevitably lose its water? Not at all, since all that is required is to lower the bowl to be siphoned so the water in the tube is, as it were, "gulped back". So, what does this all imply?

What can be done then? If we are nervous of taking the whole system into account, it would be less serious to admit that the situation is complicated and difficult to explain, rather than, with disregard for the *very principle* of the experiment, exclude a point which is embarrassing. We can simply illustrate the heart of the phenomenon, that is to say, the role played by the difference in heights between the water levels in the open air, and just leave it at that; at least consistency is not violated.

It should be noted that it is not always from a deliberate desire for simplification that truncated, or even false, explanations are produced.

The all too close similarity between teachers' typical, everyday explanations to those of their pupils leaves some room for doubt. Is this sharing the supposed "evidence"or a conscious "resonance"? By definition, these echo-explanations do not often arouse a great deal of protest or questioning among the pupils. At least initially. However, how are we to respond later to the irritation of students who, having at last been taken seriously, may object: "How come this is the first time I've been told that[111]?"

110 See Chapter 1, Figure 1.1.

111 See Chapter 6, the hot-air balloon, this comment from a student to whom the absurdity of the usual assumption has just been explained (that the pressure is the same inside and outside the canopy).

These few examples are to be taken as parables. The same ideas might be transposed without difficulty to other, slightly less simple situations: hydrodynamics, for example, is a particularly "crime-prone" area, being as complex as it is conducive to surprising demonstrations.[112] How many "BERNOULLI's principles" are there of the "less pressure, *ergo* more speed" type (or vice versa), principles remembered more for their famous author than for the conditions to which they apply, and for which generalisation verges on the absurd? And let's just pay particular attention to the pressure drop in a horizontal cylindrical tube of water, where fluid incompressibility (*just one* of the conditions for BERNOULLI's principle to apply) demands that the fluid (assumed here to be viscous, hence the principle does not apply), although at very *different pressures* at the start and end of the tube, *has the same speed.*[113] Creating the effect of surprise doesn't mean a cut-price explanation; it's better to content ourselves with the surprise, and establish clear boundaries for our ritual assertions.

These mini-demonstrations bring us closer to many of the familiar questions in popular science.

---

112 We refer the reader, for example, to the explanation provided by the Stray Cats group (Stray Cats: Lively physics & exciting experiments, ikiiki-wakuwaku demonstrations, 2005, ICPE Delhi. yoji.iida@nifty.com *et al.)* for the lift effect on a rotating golf ball.

113 However, these are parallel streamlines shown in the drawing which accompanies an "explanation" of this type: "We know that pressure and speed are related. Imagine that there is a difference in pressure between two adjacent points in the air. Because of this difference in pressure, the air experiences a force which tends to accelerate it towards a place of low pressure: In one way or another, the air is displaced by the strong pressure. In other words, the speed increases in the region where the pressure is lower. Hence we can summarise by saying that the speed is greater wherever the pressure is lower and equivalently, that the speed is lower wherever the pressure is higher", COUSTEIX J. (2001) How does an aeroplane fly? In: *Graines de Sciences,* vol. **3**, Le Pommier, Paris; BOUCHARD J.M., JASMIN D. & LÉNA P. (Eds) (in French). See also: FOURCADE S. & COLLINART P. (2008) *Les manips contre-intuitives*, Livret d'utilisation, La Maison des sciences, Paris, p. 11 "When the speed of a fluid goes up, the pressure within it goes down (BERNOULLI's principle)".

# Chapter 8

## POPULARISING PHYSICS: WHAT PLACE FOR REASONING?[114]

### 8.1 "Mission: impossible"?

That teachers and researchers have a mission to ensure that non-specialists gain a better understanding of the scientific results and the way they work in physics is beyond dispute. Journalists and editors are there to help them in how best to express themselves to a broad public. Of course, there is a tension between the need to be as precise and as complete as possible and the desire to attain a wide readership. Over the last quarter-century, a vast body of literature has been devoted to analysing this difficulty.[115]

The aim of the next few pages is simply to capitalise on the preoccupations expressed above and give some thought to the place of reasoning in popularisation.

While we have spent a considerable amount of time[116] commenting on the illusions involved in the sharing of science with the general public and on the related ambiguities and linguistic traps, the question of reasoning postulated for the "target" has been barely raised. We have spoken more about problems of inaccessibility deriving from the need for accuracy than of excessively demanding lines of reasoning. Clearly, it is accepted that the effort required of the reader, observer or listener should be reduced to a minimum. B. JURDANT expressed this position in 1975,[117] talking of popularisation and its supposed intentions: "Doing this brings no particular stress into play. Popularisation boasts that it offers painless science. Besides, this is consistent with its aim for openness...". It is hard to conceive the sharing, by the greatest number of

114 This chapter is largely a reprise of the contents of a previous article: VIENNOT L. (2007) La physique dans la culture scientifique : entre raisonnement, récit et rituels, *Aster*, n° spécial « Science et récit », **44**, 23-40.

115 For example: ROQUEPLO P. (1974) *Le partage du savoir*, Le Seuil, Paris; JEANNERET Y. (1994) *Ecrire la science*, PUF, Paris.

116 LÉVY-LEBLOND J.M. (1986) *Mettre la science en culture*, Anais, Nice; JACOBI D. (1987) *Textes et images de la vulgarisation scientifique*, Peter Lang, Berne; and ref. previous note.

117 JURDANT B. (1975) La vulgarisation scientifique, *La Recherche*, **53**, 149.

people, of certain benefits associated with science without there being an associated dimension of pleasure.

However, the need to develop a scientific culture among our pupils often goes hand in hand with a desire to develop their critical faculties. It's all about preparing these young minds for responsible citizenship in their later life.[118] If the development of a critical mind with its ability to detect inconsistency owes something to a certain rationality, and if it also contributes to the image we might have of science, then reasoning should definitely not be the pariah of the "science culture" that LÉVY-LEBLOND[119] spoke of. Pleasure and reasoning are both desirable, but are they in fact an unlikely couple?

Not all authors seem to accept this exclusion, as witness some, caught up as they were in the momentum of the "World physics year" in 2005. When clarifying the intentions of his book *Demain, la Physique* [Physics for tomorrow] (2004), E. BREZIN[120] explains thus: "The intention is to show that the questions asked are not the effect of some arbitrary factor but of an internal logic which has brought us inevitably to where we are now (...). It seemed to us that it was nonetheless possible to present the questions with which today's physicists are faced, in ordinary language and without recourse to equations or previous heavy investment of time reading difficult texts". A few years before, B. DIU,[121] had also claimed multiple objectives: "Here, he tells us, is a book on physics, truly, deliberately and passionately on real physics, with no red herrings or smokescreens. However, what is special about it (which is also its *raison d'être*) is that it attempts to present its subjects at a level accessible to the layman (a

118 General objectives set out in 1992 in France, MEN-BO Special edition of 4.9.1992, 74-75, for teaching physics and chemistry in *collège* and *lycée* (secondary school): "This teaching... seeks to develop the elements of a scientific culture among all pupils (...), it has to form critical minds (…), physics is involved in political, economic, social, and even ethical choices; physics teaching must contribute to the construction of "a 'users' handbook' for science and technology" to prepare pupils for these choices". On this subject see in particular the programme implemented in the United Kingdom to develop "scientific literacy" in a teaching programme for pupils of 15-16 years, *Twenty First Century Science*: HUNT R. & MILLAR R. 2000, Science for Public Understanding, London: Heinemann, *Twenty First Century Science*: jb56@York.ac.uk; MILLAR R. (2006) Twenty First Century Science: Insights from the design and implementation of a scientific literacy approach in school science, *International Journal of Science Education* **28** (13) 1499-1521.

119 Note 128.

120 Introduction to the work: ASPECT A., BALIAN R., BALIBAR S., BREZIN E., CABANE B., FAUVE S., KAPLAN D., LÉNA P., POIRIER J.P., PROST J. (2004) *Demain la physique*, Odile Jacob, Paris.

121 DIU B. (2000) *Traité de physique à l'usage des profanes*, Odile Jacob, Paris. A comment of Jon OGBORN concerning the adaptation of modern physics subjects to secondary school level pinpoints the associated difficulties: "This process of transposing physics at the research level into physics that can be taught in (say) secondary schools is both creative and not at all straightforward. It involves finding the right kinds of simplification, especially of the mathematical language used. It also involves a strong selection so as to present the essence of new ideas from a particular point of view." OGBORN J., *Physics education,* In J. OGBORN (Ed.) Physics now, ICPE-IUPAP, p. 72 (http://web.phys.ksu.edu/icpe/Publications).

challenge, for sure)". In the matter of science, this introduction presents the alliance between authenticity and passion with particular clarity.

The thoughts which follow relate to the implementation of programmes engaged in just such a challenge. They will build on the analyses and certain examples developed in the previous chapters in an attempt to cast light on this question: how can an author committed to such an approach increase his/her chances of making progress? And as for the "target", what seeds can be sown within the actual context of school or university training so that future readers or even active participants involved in popularisation may then be better equipped to make informed use of it?

## 8.2 Reasoning and rigour: some critical points

At the heart of many of the in-depth studies[122] lay the terminology and semiology of popular language. Metaphors, analogies and mixtures of genres imposed themselves on the analysis not only as candidates for facilitating the dissemination of science, but also as the potential culprits for all forms of semantic distortion. Here however, (and echoing the title of this chapter), we shift the focus away from questions of designation or evocation towards establishing the links between arguments produced in relation to a given question. Some elements identified in the type of text cited above will question the possible fettering of the author's ambitions, namely give the exercise of reasoning a chance.

### *The obscured question*

A marvellous experiment (in reference to FIZEAU's experiment in 1849) was carried out in the autumn of 2005 in the skies over Paris, as part of World Physics Year: a green laser beam from the Paris observatory was reflected off a mirror in Montmartre and returned to the site it came from, where a suitable device measured the duration of the journey and hence the speed of light. The "AMP Île-de-France" site, along with various other documents,[123] celebrated the experiment while explaining the principle, and indeed the apparatus itself. However, there was one question which failed to feature in any of the documents: the spectacle was marvellous, but how could this line of intense green with its sharp edges actually be seen in the Parisian sky? We will come back to this.

122 In particular see JACOBI D. (1987) *Textes et images de la vulgarisation scientifique*, Peter Lang, Berne.

123 RADVANYI P. (2006) Un rayon vert dans la nuit blanche [A green ray in the night], *Bulletin de la Société Française de Physique*, **152**, 32 ; BOBIN J.L., LEQUEUX J. & TREPS N. (2006) « C'était c à Paris » [Finding c in Paris], *Bulletin de la Société Française de Physique*, **153**, 31.

### *Explanation or tautology?*

Suppose this time that the question is posed explicitly, for example, why does the speed of light depend on the medium through which it passes. In one document written for the general public,[124] we read: "In a transparent medium, like glass for example, light propagates more slowly because its refractive index is greater than that of air." There is nothing wrong with any of the assertions in this sentence, but the word "because" linking them should not be taken at face value, since these two propositions mean the same thing.

### *Implicit links*

A favourite media topic concerning recent advances in science is that of so-called "cold" atoms. In the pamphlet already mentioned,[125] we learn that "The NOBEL Prize was awarded in 1997 to Claude COHEN TANOUDJI, Steven CHU and Bill PHILLIPS for their demonstration of the principle of laser cooling of atoms to extremely low temperatures, of the order of one thousandth of a degree above absolute zero. (...). Cold atoms are now used in atomic clocks to measure time with extraordinary precision (...)". These are entirely accurate statements, but unrelated. For a physicist it goes without saying that when talking about atoms, "cold" means having all the same low speed within the chosen reference frame. However, a layman would have more chance of understanding something if informed about the implicit link: "The interest in cold atoms is that they are slow", as we are told[126] in the previously mentioned book "*Demain, la physique*".

### *Links denied*

The risk of paralysis in reasoning is magnified when the statements presented appear to concern phenomena which are each related to a different type of modelling. Hence we read in an older text:[127]

"The pressure of the liquid on the walls (...) is fairly easy to understand (...) the molecules are squashed together (...) and push like commuters in the Metro (...). It's less easy for a gas... the molecules are no longer squeezed at all. It's when they collide with the wall and rebound that they exert pressure". In this case, the reader has some warning of a change in the explanation, allegedly "less easy" for gases. From a *repulsive* compaction model for compressible bodies in contact we pass on to a kinetically based view. Would the molecules of the liquid not cause collisions, and

124 JACQUIER B. & VANNIMENUS J. (2005) *La lumière et la matière*, EDP Sciences, Paris, p. 6.

125 Reference in previous note, p. 16.

126 Reference in note 120, p. 136

127 MAURY J.P. (1987) *L'atmosphère*, Palais de la Découverte, Hachette, Paris, 44-45.

hence so-called "kinetic" pressure? This is not the case at all, since being typically a thousand times greater than that of a gas, this kinetic pressure is not completely annihilated by the attractive forces between the molecules of a liquid. The uniqueness of the kinetic-theory model for different states of matter is here undermined, doubtless in the name of accessibility.

### The shortcut which hints at inconsistency

There is another example of a choice of writing which needs to be discussed from the point of view of establishing links. The impression of inconsistency arises this time from a shortcut leading potentially to a misunderstanding. We are talking of the "background cosmic radiation (...) which permeates space without interacting with matter", the following sentence stating that "This was discovered by accident in 1964".[128] If we understand by this (*i.e.* we already know) that the now diluted matter encountered by this background radiation affects it infinitely less than during the first moments of the universe, then all is well and good. However, in our ignorance it would be possible to interpret the first statement as that of inherent inability of this radiation to interact with any kind of matter; after all, we have heard such things said about the neutrino! What follows then, comes as a bit of a shock since it is hard to see how something which failed to interact with PENZIAS and WILSON's antenna could be discovered by accident.

These few examples illustrate the chosen written formulations for which judgment in absolute terms would be ill-advised without the dual reference, on the one hand to the author's objective, and on the other, to the reader. Assuming that the author wishes to give reasoning a chance, there is still the reader to consider. What for one is a hazardous shortcut will be easily decoded by the other, and this must be anticipated. This attitude is praiseworthy, as in the case of the present author for example who, on the media presentation of laser-cooled atoms, warns of: "(...) the seemingly paradoxical discovery (...) that judicious use of the laser can actually not warm matter, but can cool it to temperatures close to absolute zero".[129] With this allusion to a possible difficulty, the reader perceives that he is being credited with a concern for consistency, in that here we have an object capable of cutting materials and about which he must already have read something along the lines of "the laser beam is very powerful".[130] If we have this concern for the reader, it is then natural to want to have some insight into the way he thinks.

---

128 DIU B. (2000) *Traité de physique à l'usage des profanes*, Odile Jacob, Paris, p. 280.

129 ASPECT A., BALIAN R., BALIBAR S., BREZIN E., CABANE B., FAUVE S., KAPLAN D., LÉNA P., POIRIER J.P., PROST J. (2004) *Demain la physique*, Odile Jacob, Paris, p. 134.

130 JACQUIER B. & VANNIMENUS J. (2005) *La lumière et la matière*, EDP Sciences, Paris, p. 4.

## 8.3 Stories – the layman's penchant

A major determinant of a non-specialist's reading of a message relating to science has long been identified. The "common knowledge" as analysed by BACHELARD[131] has been one of the source topics of the educational studies over the last three decades with regards to students' "conceptions", *i.e.* very widespread views about scientific concepts and/or the nature of science. For example, we have already identified a common tendency for excessive transformation of concepts into things, typically a ray of light being understood as an object visible from everywhere. Another illustration of this common approach is that of an optical image, understood as travelling as a single entity from the source to the other elements over a preferably horizontal path up to the screen on which it is displayed.[132] It is reasonable to say that such ways of thinking act as filters, selecting in some sense that part of the message consistent with the receiver's personal passband, or (using a less unidirectional metaphor), they determine particular resonances.

We may assume that this applies broadly to a major tendency in common reasoning, causal linear reasoning.[133] As a general rule, this preferred reasoning structure among "laymen" is related to what they understand from a science-related message.[134] The effect is without doubt all the stronger when the author himself finds some sort of resonance with the characteristics of common reasoning. This is frequently found among authors who wish to get themselves understood "painlessly"(to use B. JURDANT's expression (ibid.)), pleasurably even, by giving the listener or reader the impression that thanks to the story they have understood the essential. Clearly, the "echo-explanation" already illustrated in Chapter 7 finds fertile ground in popularisation.

131 BACHELARD G. (1938) *The Formation of the Scientific Mind.* (Translation Mary McAllester-Jones) Manchester: Clynamen Press, 2002.

132 GALILI I. & HAZAN A. (2000) Learners' knowledge in optics: interpretation, structure, and analysis, *International Journal of Science Education*, **22** (1) 57-88; VIENNOT L. (2001) *Reasoning in Physics - The part of common sense*, Dordrecht: Kluwer Ac. Pub., Chapter 2.

133 See this work, Appendix C.

134 In VIENNOT L. (2001) *Reasoning in Physics - The part of common sense*, Dordrecht: Kluwer Ac. Pub., Chapter 5, 112-114, an example of an "explanation" for changes of state of the form: At the start, gas → temperature falls → kinetic energy falls → molecules can no longer resist the interactions → they agglomerate into the liquid state → then into the sold state. 77% of the first year students consulted (N = 181) said they understood that at equilibrium the mean kinetic energy per particle is lower in the liquid than in the solid phase, which is not correct. Hypothesis: chronological reading of the arguments suggests this idea (except for subsequent precaution, actually present in the work in question): at the start, gas, at the end, liquid, meanwhile, the drop in temperature and that of the kinetic energy, which is necessarily lower at the "end", *i.e.* in the liquid.

## 8.4 Authors (as well as teachers): the attraction of "echo-explanation"

### *In tune with habitual modes of reasoning: oversimplified stories*

Thus, it can be said that, habitual modes of reasoning happily inspire a popularising style. But we, the guardians of knowledge, whose role we assume as creators of material for the general public are in ourselves not immune to certain tendencies. In order to attain better control of the final impact, we need to be aware of these.

As far as everyday modes of reasoning is concerned, and hence what probably characterises "laymen", one of the most often observed is their tendency to misunderstand concepts as ordinary objects. The authors certainly don't say that light is seen in profile like a passing train. However, not to say how the green beam which pierced the Parisian sky in the autumn of 2005,[135] became visible, isn't that like saying you could see it because it was there?

Reasoning using just a single variable is a precious and often over-used resource. In one text,[136] since revised, we read: "Aircraft fly very high, at an altitude where the molecules are far less numerous and hence the external air pressure on the windows is much less than at sea level". A few pages further on, talking about a hot-air balloon, it is explained that the effect of heating the internal air was such that it contained "less and less air". Simple consistency then suggests that the air is less dense and hence that the pressure is lower, a fatal assumption for keeping a balloon afloat. Of course, the proposed explanation then veered off in another direction, referring to ARCHIMEDES' principle. Regarding jets of water emerging from a pierced bottle, Appendix D contains another illustration of the dangers of an analysis reduced to a single variable. Consistency often seems to be parenthesised, in favour of blow by blow explanations somewhat similar to those frequently observed among pupils.

Finally, and most importantly, the chronological structure of everyday explanations can be observed in many attempts to popularise. The explanation provided for the siphon mentioned in the previous chapter[137] has already neatly illustrated this point. Starting from the free air end of the water-filled tube, as an explanation of what is happening, then invoking a vacuum somewhere in the fluid, and then going to see what is happening on the side of the bowl, is tantamount to denying the essential feature: in order to predict the behaviour of each of its parts, the whole system must be

---

135 See note 123.

136 MAURY J.P. (1989) *La glace et la vapeur, qu'est-ce que c'est ?* [Ice and steam: what are they?] Palais de la Découverte, Paris, p. 27.

137 CHAVANNES I. (2003) *1907. Leçons de Marie CURIE aux enfants de nos amis* [Marie CURIE's lessons to our friends' children], EDP Sciences, Paris, p. 62.

taken into account simultaneously. The local analysis pointed out by FAUCONNET[138] for its irrelevance, is here manifest again.

Difficulty in taking duration into account can be clearly seen in everyday explanations of the steady-state; thus it is for the greenhouse.[139] How many times do we find that the "explanation" for a greenhouse is that solar radiation enters, only to find itself "trapped" by such and such a phenomenon, and so less radiation escapes than comes in? In a schematic sense, *radiation enters → part of it is trapped → the inside temperature goes up*. Most of the time it is not made clear that there is a regime in which the temperature is rising, but that there exist steady or quasi-steady regimes for this phenomenon with flows of energy for which the input and output powers are (nearly) equal. An internet search produces cataclysmic images in which the imbalance between the radiation input and output would suggest a disastrous explosion of the planet if, that is, we ascribed some permanent validity to the image, during the nightime as well?[140] For a lettuce growing in a greenhouse at a stable, or even decreasing temperature[141] (at night for example), we can no longer say that more energy is coming in than escaping.

Here is another example of an explanation in the form of a story. An author already mentioned[142] talks of "the cosmic background radiation, left behind when the universe expanded ten or so billion years ago and which has since bathed interstellar space without interacting with matter". The chronological structure of this passage comprises three episodes: *Radiation interacts with matter → the expansion abandoned it → it permeates space (from then on)*. How can the layman perceive the steadiness of the expansion and the continuous cooling of the radiation? Will he understand that the 2.7 °K currently measured is not actually the end of the story? The effect of this text has not been assessed, and evidently depends on the public. We simply point out that the question deserves attention.

---

138 FAUCONNET S. (1981) *Étude de résolution de problèmes : quelques problèmes de même structure en physique*, Third year thesis, LDPES, Université Paris 7. And reference in the following note, 95-99.

139 VIENNOT L. (2001) *Reasoning in Physics - The part of common sense*, Dordrecht: Kluwer Ac. Pub., 110-112.

140 See for example: http://www.cea.fr/jeunes/mediatheque/animations-flash/energies/energie-et-effet-de-serre; and also official documents: National Ministry for Education, Research and Technology [Ministère de l'Education Nationale, de la Recherche et de la Technologie]. The greenhouse effect. In P*rogrammes et Accompagnement du cycle central (5ème, 4ème)*, Paris, CNDP, 92-93, 1998, re-ed. 1999. More details can be found in COLIN P. (2011) Enseignement et vulgarisation scientifique : une frontière en cours d'effacement ? Une étude de cas autour de l'effet de serre, *Spirale*, 48, 63-84.

141 Even if the internal temperature remains higher than the external temperature.

142 DIU B. (2000) *Traité de physique à l'usage des profanes*, Odile Jacob, Paris, p. 280.

### *The lasting effect of teaching rituals*

Other, less subtle, effects are liable to affect discursive speech of science reporting. This has to do with habits acquired at school or university, reflections of our own teachers. There are a few more examples. It is part of the ritual (and besides, with no serious consequence, it has to be said) to assert that diffraction is observed for apertures whose diameter is "close to the wavelength".[143] Just imagine (perhaps we shouldn't *even think about it*?) the effort required in practicals if we had to use slits of the order of a micron! Fortunately, a factor of a thousand gives more reasonable levels of luminous flux, while still permitting the investigation of diffraction.

Another refrain, which has, however, been castigated for a long time,[144] is that "In optics, frequency is equivalent to colour".[145] We even go on to talk of "colours invisible to the naked eye",[146] even though colour is a *perceptual* response to the light received. As for colours which are by definition "visible", there is no one-to-one correspondence between colour and frequency. Thus, we can see yellow without there being any "yellow frequency" in the spectrum of the received light,[147] even though the famous sodium doublet is well and truly yellow. As for finding a magenta frequency....

In the matter of ritual, we also remember the "isobaric" hot-air balloon discussed in Chapter 6. No surprises there in spotting it in a popularising context![148]

## 8.5 A real margin for manœuvre

Of course, we could say that more rigour would not be a bad thing, but on the other hand, what do we do about the spectre of an abstruse and boring physics looming on the horizon. Besides that, is making an approximation here and there really so serious? In view of the limitations inherent in popularisation, is the margin for manœuvre so small that this very question becomes irrelevant?

---

143 Nicolle J. (2005) Lumière sur la lumière, *Réflexiences*, **2**, 6.

144 Chauvet F. (1996) Teaching colour: designing and evaluation of a sequence, *European Journal of Teacher Education*, **19**, n°2, 119-134. See also: (http://www.lar.univ-paris-diderot.fr/sttis_p7/color_sequence/page_mere_fr.htm); Planinšič G. (2004) Color light mixer for every student, *The Physics Teacher*, (**42**) 138-142; Planinšič G. & Viennot L. (2010) *Shadows: stories of light* (www.eps.org, select *Education* and then select *MUSE*).

145 Diu B. (2000) *Traité de physique à l'usage des profanes*, Odile Jacob, Paris, p. 242.

146 Maarek S. (2002): (http://www.fondation-lamap.org/fr/page/11231/spectre-de-la-lumi-re-blanche).

147 This is the case when, for example, two laser beams, one red, one green, are superimposed on a screen wich is white in white light.

148 Maury J.P. (1987) *L'atmosphère*, Palais de la Découverte, Hachette, Paris, p. 67.

The examples above are given to underline the choices which are indeed open. They provide elements on which to meditate if we wish to give the reader, listener or participant a chance for reasoning, while, at the same time, not drowning him in a flood of formal or technical details. Recent experience with a small group of future scientific journalists showed them to be very open to such an approach.[149]

We can *decide* to pinpoint the questions which come up (why do we see the laser beam?) even without responding to them;[150] to clarify the links which our professional jargon disguises (those "cold" atoms, whose interest for us is that they are "slow"); to state the supposedly obvious, such as the continuous expansion of the universe; to recognise the surprising, such as the fact that these lasers, usually synonymous with power, can also play their share in cooling; to acknowledge that certain phenomena are not determined by a single variable, even though so many statements wrongly suggest that they are. We can decide on all these points all the more lucidly if we are aware of our own inherent tendencies; simple humility impels us to acknowledge that they are probably very widely shared.

Above all, we can learn to avoid putting all these liberties taken with the physical theory on the same level. Certain "details", it is true, can be passed by in silence, and certain inexactitudes deliberately incorporated in the chosen models provided that the essential part of the message remains intact. The situation, however, is completely different, for example, with the "isobaric" hot-air balloon, where the very principle of the phenomenon was written off. If he is uncertain of how to *order* the different facilities that he has at his disposal to present science to the greatest number,[151] (this being related to their effect on understanding), an author is in danger of pursuing the illusion of accessibility, whereas in fact he is completely obscuring his message. The reader or listener must actively participate in this enterprise if they are to avoid taking erroneous information at face value, otherwise, the most vigilant may feel frustrated, confronted with what surely appears to be inconsistency.

For all that, the difficulty cannot be denied. There is a good reason why the alliance of reason with pleasure seems somewhat against nature, even if the term "pleasure" might be renamed in a more specific and more appropriate manner, viz. "intellectual

149 Mathé S. & Viennot L. (2009) Stressing the coherence of physics: Students journalists' and science mediators' reactions, *Problems of education in the 21st century*, **11** (11) 104-128. See Appendix E.

150 To be specific, Rayleigh scattering by air molecules plays a crucial role in the sharpness of the observed green beam (this frequency is more favourable for this phenomenon than that of a red laser beam); this sharpness is poorly explained by "scattering by dust" alone, as ritually explained.

151 This is a topic introduced and explored at Year 11 (*Seconde*) level by Ivan Feller in his 2008 thesis: *Usage scolaire de documents d'origine non scolaire en sciences physiques. Eléments pour un état des lieux et étude d'impact d'un accompagnement ciblé en classe de seconde,* Université Paris Diderot (Paris 7). See Appendix F and Feller I., Colin P. & Viennot L. (2009) Critical analysis of popularisation documents in the physics classroom. An action-research in grade 10, *Problems of education in the 21st century*, **17**, 72-96.

satisfaction". Success in this domain is not achieved without a great deal of attention. There is a long list of supposedly illuminating "mini-demos", (see Chap. 7), high-impact images or vibrant explanations which actually turn out to be barriers to reasoning. Since it's a matter of choice here, we can–without irony–do this to favour illumination, shock and emotional buzz. However, if we are aiming to give the "layman" the satisfaction of feeling that he is engaged in reasoning and for him to enjoy the fruits of so doing,[152] it is essential to be quite aware of the somewhat technical aspects discussed above. Of course, providing a narrative form of explanation always engenders a comfortable feeling of familiarity; this can, at times, be appropriate, as witness these cosmic tales which astrophysicists have put to good use. For all that, we should keep an eye out for the possible consequences (in terms of consistency) of explanatory scenarios which plainly echo the everyday tendencies of reasoning, in particular those in an exclusively causal linear framework.

152 Viennot L. (2006) Teaching rituals and students' intellectual satisfaction, *Phys. Educ.*, **41**, 400-408 (http://stacks.iop.org/0031-9120/41/400); Viennot L. (2005) Les valeurs de la science, *Science et Avenir*, Hors série, 144.

# Chapter 9

# Conclusion

As we said at the beginning, our ambition in writing this text was to contribute to the universally acknowledged desire to promote understanding and reasoning in physics. The level chosen is relatively elementary. The examples put forward may cast light on the processes of popularisation and teaching in physics from the very start until the first years at university level. The guiding ideas aim to be more generally applicable.

What are these guiding ideas?

The starting point is to emphasise one particular characteristic of physics, namely the elegance and parsimony of physical theories: just a handful of laws covering numerous phenomena.

Enjoyment of such an intellectual conquest presupposes that due respect is paid, with a concomitant regard for internal consistency of the discipline. In this context we cannot state a law (such as one of NEWTON's laws) appropriate to its domain of application and then violate it an instant later, for example, by announcing the existence of non-equilibrium forces to justify some mechanical equilibrium, or denying the fact that any interaction associates two opposing forces. Positively (this time) the desire to emphasise the power of the functional approach in relationships follows, the desire to bring out the links between the analysis of phenomena which at first sight seem very diverse, and also the desire to emphasise the convergence of different approaches to the same phenomenon, by, for example, the scale of the description.

On such topics, the examples used here are simple. An inverted glass of water serves to highlight the idea (devastating as far as Newtonian consistency is concerned) that an object always exerts its own weight on the surface supporting it. It's another glass on a weighing machine that emphasises the often ignored reciprocal nature of the interaction between the water it contains and an immersed ball. A siphon, a "love-meter" and a duck which nods ceaselessly remind us that it's differences that "make the world go round". As regards a unifying formalism, a simple proportional relationship suffices to deal with phenomena as diverse as the time difference between the light and sound received after a lightning discharge, a bat's obstacle avoidance system, and RÖMER's discovery that the speed of light is finite.

This is precisely the occasion to put this appearance of simplicity into perspective. Especially (but not uniquely) when popularising, the presumed specialist who expounds on a "simple" example runs a serious risk of acting without due care; there are two reasons for this. One is that a listener who knows little about physics is also deemed to be uncritical. The other is that our rituals have caused us to lose any sense of danger, hence the stress that we place on analysing certain seemingly trivial subjects in greater depth. But make no mistake: this isn't just to avoid errors in reasoning, such as the simplistic claim that a hot-air balloon can have the same pressure on either side of its envelope and remain undamaged. The aim is that the reader or listener should, if possible, gain greater intellectual satisfaction.

But, "Is it possible?" The whole question lies there. We know how tempted teachers are to be fatalistic working under so many constraints... In claiming as it does to have a positive effect on the *practice* of teaching, what does this book actually bring to the reader?

This has been discussed already: as far as great principles are concerned, there is nothing new here. However, if big ideas are not involved, what about the little ones? We can reply in at least two ways.

One is that the reader adopts two or three suggestions to create a personal set of teaching tools, for instance those optical conjugate hyperbolae we never thought about, that so-called "isobaric" balloon that merits greater attention, or that upside-down glass which works just as well when it's horizontal. Or else: when students are working on an answer sheet, with all the calculations provided, taking the opportunity to refocus on the *meaning* of the suggested elements of the solution and on their potential to tackle the related physical situations.

Another way of reading the book is to see in it the trace of a process. Any imperfections in the suggestions then become of secondary importance for, if the reader happens to surpass the author in her own field, then this is the proof that a view of teaching has been shared.

Great ideas often progress through well thought-out details, and conversely, their neglect can be the cause of failure. We can have a lot of sympathy for such and such a "method of teaching", but, regardless of what it is, a method is of no use in teaching unless considerable attention is paid to everything it leaves undetermined. "The devil is in the detail" as the adage goes, but that is not the only thing; the pleasure of thought feeds on it too.

Details, certainly, are not always the most important issue and this explains why certain aspects of physical phenomena can be ignored. Either explicitly or implicitly, under cover of a model or otherwise, we follow the teachers' widely accepted words: "We can't say everything all at once!" This is all the more legitimate if the omissions in question do not erode the essence of the message or its internal consistency. Popularisation is not alone in falling back on this assessment; it underpins teaching as well.

Both positively and negatively, this book emphasises the finer aspects of teaching, *i.e.* assertions, practical sessions and omissions, all of which are critical, in so much as potentially they may result in intellectual, not to say affective, consequences. And this raises further questions.

Concerning the possible negative consequences of carelessly prepared practical sessions (*i.e.* the proverbial "devil"): is it important to leave somebody believe something false, especially when nothing which is false has actually been said?

On the positive side, there are the advantages which come directly from taking both physics and students seriously: when a topic has been tackled carefully, what value can we attribute to these moments of enthusiasm if we cannot do the same for other topics.

It is up to each teacher or populariser to assess the response he or she wishes to give to these questions, according to the perceived constraints, risks and potential benefits.

In any event, all those studies which have so abundantly described the risks deriving from students' common difficulties cannot be ignored. That being the case, it is the responsibility of each expert who wishes to share a little physics with others who are less advanced, to measure the probable consequences of his choices. Paradoxically, there is one phenomenon which tends to obstruct this appraisal: conditioned by habit and set in our rituals, we deliberately turn to common sense in our explanations, thereby reinforcing the related risks. Reasoning using a single variable is given full rein. All too frequently our way of handling the "attractive" experiment deliberately maintains the illusion that we are directly seeing the concept, and the unstated implications in the images are not kept in perspective. From this point of view, it is perfectly reasonable to give some thought to our tendency to use "echo-explanations", those explanations which resonate with common sense. Hence, there is much to be gained from a simple examination of the structure of our arguments, which so often rely on sequences of binary causal determinations (*one* cause, *one* effect), making them seem like the narrative of children's stories.

Some questions still remain concerning the potential benefits: these have, in fact, to be experienced to be believed. At minimal cost, whether in teaching time or equipment, signs of intellectual pleasure among students can be seen appearing: this is intensely gratifying for the teacher! But how can we measure such effects more rationally, or assess the long-term consequences? This is not simple. Very few research studies tackle the question of the pleasure students take in reasoning and in witnessing their own intellectual progress. In research investigations the affective aspects are often seen as *conditions* for learning or recruitment;[153] the pleasure of reasoning and understanding is not seen in itself as a highly desirable *product* of the teaching process. In this respect, several results from the surveys cited here (though still patchy) provide an encouraging view of what is possible by way of actions which are both marginal and decisive. At the very least, they endorse this observation: in using the arguments on which this book is based, we have been able to evoke among not particularly brilliant school pupils or students such magnificent comments as: "You've made me think, thanks". We hope that all the efforts made in teaching and the media to make physics in particular, and science in general attractive do not lose sight of the very profound satisfaction that anyone can experience in the exercise of reasoning to gain a little extra understanding.

There remains a great deal of room for manoeuvre in this fine training objective.

---

153 Laukenmann M., Bleicher M., Fuller S., Gläser-Zikuda M., Mayring P. & Rhöneck C.V. (2003) An investigation of the influence of emotional factors on learning in physics instruction. *International Journal of Science Education*, **25** (4) 489-507; Pintrich P.R., Marx R.W. & Boyle R. A. (1993) Beyond cold conceptual change: The role of motivational beliefs and classroom contextual factors in the process of conceptual change, *Review of Educational Research*, **63** (2) 167-199; Rhöneck C.V., Grob K., Schnaitmann G.W. & Völker B. (1998) Learning in basic electricity: how do motivation, cognitive and classroom climate factors influence achievement in physics? *International Journal of Science Education*, **20** (5) 551-565.

# Part IV

## Appendices

# Appendix A

## WHAT THIS BOOK OWES TO PHYSICS EDUCATION RESEARCH

The suggestions and discussions gathered here are, to a greater or lesser extent, rooted in research into teaching and learning. We mention certain results when they seem particularly relevant. However, even when we may not be referring strictly to formal research the basic principles of such practice are never too far off. Let us briefly put all this in context.

Fortunately, researchers are not alone in highlighting the advantages to teachers of having a clear idea of both the extent of students' previous knowledge and their ways of reasoning, *before* starting to teach. This might seem self-evident, but it is only recently that the importance and relative consistency of widespread ideas we thought our students had not been taught have been measured. "Intuitive concepts", "naïve ideas", "common-sense ideas" are some of the labels used in this connection following the massive development of work in this field since the late 1970s.[154] These "common ideas" are widely shared and extremely resistant to teaching and the first reaction in the research community was to try to oppose these ideas, a position that was nuanced later on. Be that as it may, these findings fitted in with the notion of considering each student as having an essential role to play in the development of his/her own understanding. We note that adopting this position does not necessarily imply a radical form of "constructivism" which might end up marginalising the teacher. However, the fact remains that during the 1980s attention was increasingly focused on the student. It is only recently (roughly fifteen years ago) that the main participant in the teaching process, *i.e.* the teacher, regained as central a role as the student in the educational process. The attention given to the course content has not been eclipsed in these thirty or so years. By definition, physics education research focuses on the understanding of aspects relating to the precise content of a given discipline, even though other more transversal topics might come into the discussion. Clearly, focusing on the content involves serious restrictions, with the result that, despite their crucial importance, the sociological and psychological aspects of

---

154 A bibliography by Reinders DUIT (formerly Helga PFUNDT & Reinders DUIT) includes about 8,400 entries: DUIT R. (2006) *Students' and Teachers' Conceptions and Science Education* (http://www.ipn.uni-kiel.de/aktuell/stcse/stcse.html).

educational situations are forced into the background. This specialism claimed by physics education research requires researchers to have an in-depth knowledge of the content in question. Rather than "knowledge", it may be better to talk of the "laborious humility" we have to adopt when dealing with content whose familiarity often masks its subtleties.

This book is aimed at teachers, and what interests us above all are certain facts about teaching from which we can learn. However, what exactly is a "fact" in this context? Over the hum of conversation in the staff canteen, teachers can often be heard to say: "I did that, and then such and such...". This is then followed by some objective facts, average marks and resulting opinions. What research is aiming at, as far as "facts" are concerned, is more complicated. If it's a teaching method we want to assess, for example, we need to probe several different areas: the conceptual goals, the expectations of the researcher in terms of the obstacles as well as the potential pathways to learning, the teaching procedures, the indicators used to assess what occurs over the course of implementing the method against this background, and what the results were, even several weeks later. Ideally, qualitative indicators such as class dialogue, and quantitative indicators such as tests and marking, should both be used to throw light on the process. It is especially useful to assess areas of potential loss and gain. Faced with such difficulties, we cannot always guarantee great consistency in an assessment, though this is the goal for which researchers aim.

We are discussing here a number of experiments which differ widely in terms of their assessment. Where appropriate, the "facts" put forward are clearly indicated as research, and not simply as personal (and therefore anecdotal) tries that may, or may not, be conclusive.

# Appendix B

## The weight of air and molecular impacts: how do they relate?

*This appendix reproduces in its near entirety an article of the same title published in the Bulletin de l'Union des Physiciens (2010), vol. 104, n° 922, 263-268, reproduced here with the kind permission of the Publisher.*

This note suggests and discusses responses to the question: how does the pressure due to molecular impacts on the ground correspond exactly to the Newtonian force balance? In other words, the value obtained when the weight of a column of air is divided by the area of its base.

A recurring question, in particular among trainee teachers, conveys the unease which the topic of atmospheric weight generates: how do the molecules which strike the surface of the ground "know" that their impacts must pass on the weight of the column of air above them?

The replies suggested here are aimed only at feeding a discussion on this subject. If they can help to dispel the impression of mystery or help to identify it better, then they will have attained their goal.

After a brief description of the classical treatment of an isothermal atmosphere, a more intriguing formulation of the result associated with this everyday course question is proposed. This is followed by a chain of reasoning which contains nothing unprecedented beyond the fact that it draws together a particulate view and gravitational considerations. As far as possible, this presentation aims at simplicity.

### B1 Classical analysis of an isothermal column of air

The standard way in which an isothermal column of air (of base area $dS$) is analysed is summarised below. Here this involves no more than the practice of common physics, and we know that a presentation consistent with a mathematical treatment involving differentials is entirely possible.[155]

A volume of atmospheric air between altitudes $z$ and $z + dz$ (Fig. B1) is subjected to two unequal forces in opposite directions acting on the horizontal faces located at these altitudes. The resultant of the other forces exerted by the air on the layer in question (horizontal forces) is zero.

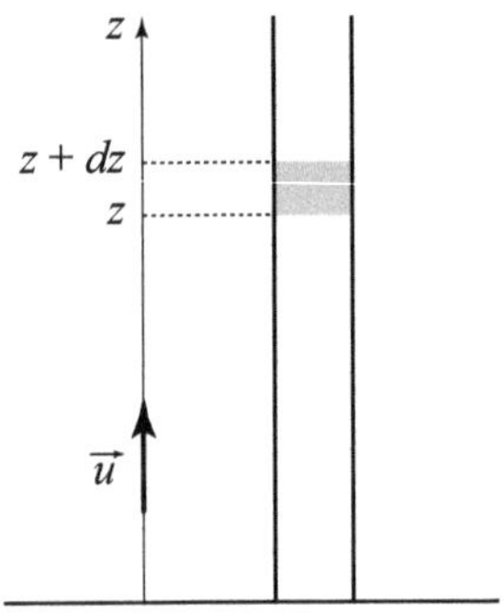


***Figure B1*** *- Diagram for a column of atmospheric air*

The resultant force $d^2\vec{F}$ exerted by the outside air on the slice is thus: $d^2\vec{F} = [p(z) - p(z+dz)]\, dS\, \vec{u}$ where $\vec{u}$ is an upwards directed vertical unit vector.

This equality may be written: $d^2\vec{F} = -\,dp\, dS\, \vec{u}$

Moreover, the weight of the slice $d^2\vec{P}$ is $d^2\vec{P} = -\,g\rho(z)\, dz\, dS\, \vec{u}$ where $\rho(z)$ is the density of the air at that altitude.

At equilibrium, the resultant of all the forces exerted on the slice is zero:

$$d^2\vec{P} + d^2\vec{F} = \vec{0}$$

and hence
$$g\rho(z)\, dz\, dS\, \vec{u} + dp\, dS\, \vec{u} = \vec{0} \qquad (1)$$

*i.e.*
$$\frac{dp}{dz} = -\,g\rho(z) \qquad (2)$$

Finally, the ideal gas relation ($pV = nRT$, in the usual notation), assumed applicable here, leads to $\rho = \dfrac{Mp}{RT}$ where $M$ is the mean molar mass of the air.

Hence
$$\frac{dp}{p} = -\frac{Mg}{RT}\,dz \qquad (3)$$

And in an isothermal atmosphere,

$$p(z) = p_0 \exp\left(-\frac{Mgz}{RT}\right) \text{ where } p_0 \text{ is the pressure at altitude } z = 0. \qquad (4)$$

155 Artigue M., Menigaux J. & Viennot L. (1990) Some aspects of students, conceptions and difficulties about differentials, *European Journal of Physics*, **11**, 262-267.

The force exerted by the column on the ground is $-\, p_0\, dS\, \vec{u}$. Taking account of NEWTON's third law, a global balance of forces on the column allows us to say that the weight $d\vec{P}$ of the column is also $d\vec{P} = -\, p_0\, dS\, \vec{u}$

The weight $\vec{P}$ of the column may also be found by integration:

$$d\vec{P} = -\int_0^z g\rho\, dz\, dS\, \vec{u}$$

*i.e.*

$$d\vec{P} = -\int_0^z g\frac{Mp}{RT}dz\;\; dS\,\vec{u} = -\, g\frac{MdS}{RT}\int_0^z pdz\,\vec{u} \qquad (5)$$

And since

$$\int_0^z pdz = p_0\frac{RT}{mg} \qquad (6)$$

then

$$d\vec{P} = -\, p_0\, dS\, \vec{u} \qquad (7)$$

In other words, the product of the pressure of the column of air at equilibrium at ground level with the contact area between ground and column is equal to the weight of the column.

## B2 Another way of looking at things

And that could be the end of it all, with no further questions asked. A global Newtonian force balance for an atmosphere in equilibrium is an unquestionable argument, and the integral confirms the relation between weight and pressure on the ground. However, as is often the case, a formal argument of the type "it is necessary that..." is not at all satisfying. Those who are unhappy with it (and often know much more than is imagined) sometimes make an equivalent reformulation:

*The impacts of the molecules on the ground result in a force equal to the weight of all the molecules above, as if they were touching each other in the static situation.*

This result–irrevocable, of course–is more thought-provoking. It leads us to take another line of approach, a more local line of reasoning, and this is the purpose of the following proposition.

## B3 Molecules, impacts and weight: a proposal for their analysis

In a cylindrical box with horizontal ends at heights $z$, $z + \Delta z$, a particle of mass $m$ (the only one occupying the otherwise empty space) goes back and forth between the two ends. The moduli of its vertical velocities at the lower and upper altitudes respectively are $v$ and $v + \Delta v$ on account of gravity. The time taken to fall, as the time taken to rise, is $\Delta t$. At each end, the effect of the impact is to "return" the momentum.

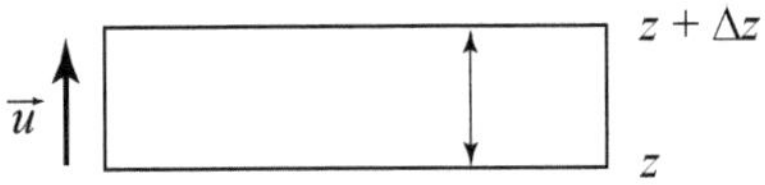


***Figure B2*** - Can we show that the action of this particle on the container is on average equal to the weight of the particle?

As regards the weight of the atmosphere, we can see at once that this result immediately appears more problematical than its global equivalent. During a survey among degree-level students, 12 out of 13 thought this result inexact for one molecule, even though the relationship between the weight of the atmosphere and the pressure on the ground had at first seemed to be perfectly clear. Let us see a way to convince ourselves of this.

At a given instant, the force exerted by the container on the particle obeys NEWTON's second law $\vec{f} = m\vec{a} = \frac{d\vec{\boldsymbol{p}}}{dt}$ where $\vec{\boldsymbol{p}}$ is the momentum of the particle (here in bold to avoid confusion with pressure).

During a period $T$, the average value of this force is

$$\vec{F}\ ave.\ (container\ on\ ptle.) = \left(\frac{1}{T}\right)\int_0^T \vec{f}\,dt$$

$$= \left(\frac{1}{T}\right)\int_0^T \overrightarrow{d\boldsymbol{p}} = \frac{\Delta\vec{\boldsymbol{p}}}{T} \qquad (8)$$

$\Delta\vec{\boldsymbol{p}}$ is the change in momentum due to the container over the period $T$.

Consider a period $T = 2\ \Delta t$, *i.e.* a two-way trip for the particle. Two impacts took place at speeds $v$ and $v + \Delta v$ respectively.

- Variation in momentum $\vec{\boldsymbol{p}}$ for an impact at the top of the container:

$$-2m(v + \Delta v)\ \vec{u}$$

- Variation in momentum for an impact at the bottom of the container:

$$2m(v)\ \vec{u}$$

- Variation in momentum *due to the container* for a two-way trip:

$$\Delta\vec{\boldsymbol{p}}\ = -2m(\Delta v)\ \vec{u}$$

The increase in speed between top and bottom is determined by the weight in free fall: $\Delta\vec{v} = \vec{g}\Delta t$, *i.e.* $\Delta v\,\vec{u} = -\,g\ \Delta t\ \vec{u}$

Hence $$\Delta\vec{\boldsymbol{p}} = 2m\ g\ \Delta t\ \vec{u}$$

and from (8) : $$\vec{F}\ ave.\ (container\ on\ ptle.) = 2mg\frac{\Delta t}{2\Delta t}\vec{u}$$

*i.e.* $$\vec{F}\ ave.\ (container\ on\ ptle.) = mg\ \vec{u} \qquad (9)$$

The force exerted by the particle on the container is therefore, on average, $-mg\,\vec{u}$, *i.e.* the weight of the particle. Via the impacts on the walls of the container, the particle exerts, on average, a force equal to its weight.

This line of reasoning is valid regardless of the value of the speed $v$.

It applies just as well to the vertical component of any non-vertical particle velocity.

The impacts occurring between particles conserve momentum. The fact that these may occur over the course of the period being considered does not alter the time-average of the forces exerted by the molecules on the walls or on the ground.

Finally, in the absence of horizontal walls, the transfer of momentum from the adjacent layers to the layer being considered is the same as that due to the imaginary walls being considered here. Classically speaking, we make no distinction between the pressure on a wall and that existing within a fluid at equilibrium.

This line of reasoning therefore meets the well justified result: via the impacts, the action of the molecules on the ground is the same as if all the molecules in the atmosphere were all piled up, stationary on the ground.

## B4 Final remarks

Although we may agree that a source of explanations can be found via the approach above, we still need to know in what way and at what level our students can benefit from it. We may be able to spot certain positive aspects, or on the contrary, potential obstacles to comprehension.

First of all, we may wonder how the global argument "the centre of inertia theorem must be satisfied" can be true at the local level of a molecule. Other work has shown the advantages of such changes of scale in the analysis. Besides making it more comprehensible, we achieve a type of *link-making* procedure which emphasises all the power and elegance of a physical theory, *i.e.* Newtonian mechanics ~~is~~ supported here by the kinetic theory of gases. Moreover, it is not only the global and local approaches which are reconciled here, but also a static and dynamic view.

For all that, the difficulty should not be underestimated, especially that resulting from the use of averages. This is a potential source of confusion, since justifiably the individual aspect (the analysis of the dynamics of a *single* molecule) becomes somewhat dilute.

Note that thermodynamics comes in only by the "back door". As we might expect, it does not appear so long as just one molecule is involved. To begin with, it is less clear that the distribution of speeds does not affect the conclusion established with one molecule in a slice, since this conclusion does not depend on speed at all. In the same purely Newtonian line of analysis, it can also be shown that a steadily flowing inverted hourglass (there is still sand at the top, and already sand at the bottom) exerts the same force on a weighing machine as when the situation is static, with no flow. This example helps moreover to demystify the question of the weight of the atmosphere.

The perfect gas relation makes an appearance only when it is necessary to correlate the density and pressure and to perform the integration. Discreetly, but concealing what remains mysterious, thermodynamics tells us that in an isothermal situation it is the number of particles which determines the (exponentially decreasing) variation in the mutual forces between slices: there are more particles near the ground than higher up. We have to admit that attempts at mechanistic reasoning (beyond the usual "it is necessary that...") come up against this level somewhat, in imagining how the molecules "know", this time, where and in what proportion to gather together. BOLTZMANN had to get involved.[156]

**156** For a discussion on this point, consult in particular the article by DEVAUD M. & TREINER J. (2010) Théorie cinétique de la pesée d'un gaz, *Bulletin de l'UDPPC*, **104** (928), 1021-1024; also VIENNOT L. (2011) Le poids de l'air, le choc des molécules : il fallait bien que BOLTZMANN s'en mêle, *Bulletin de l'UDPPC*, **105** (932) 313-315.

## B5 The weight of molecules: a survey among trainee teachers

The following questionnaire was put to a group of trainee teachers (second year at teachers' training college [*IUFM*], PLC2, N = 19) :

In a cylindrical container with horizontal ends at heights $z$, $z+\Delta z$, a particle of mass < (the only one occupying the otherwise empty interior) goes to and fro under the action of gravity and collisions on the walls* over a vertical path between these two ends.

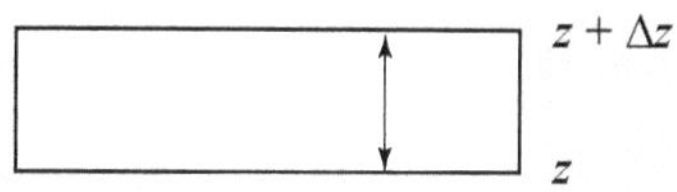


- *Question 1*

Is the following statement in italics true?

*The action of this particle on the container is, when averaged over time, equal to the weight of the particle.*

□ yes □ no □ dont' know

Explain your answer:

- *Question 2*

Is the following statement in italics true?

*The action of the molecules on the ground, via their collisions, is the same as if all the molecules in the atmosphere were all piled up, stationary on the ground*

□ yes □ no □ dont' know

Explain your answer:

- *Question 3*

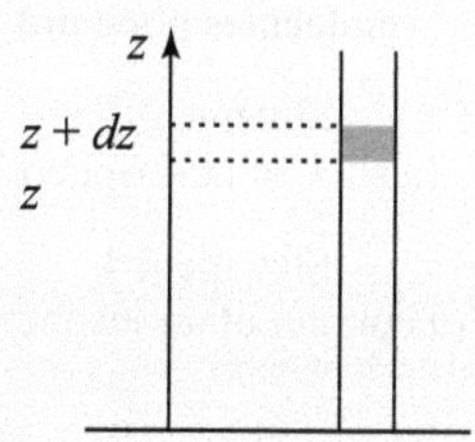


***Figure 1*** *- Diagram for a column of atmospheric air*

When analysing a column of atmospheric air, it is normal to write that:

*For a slice at equilibrium between altitudes z and z+dz (Fig. 1), the resultant force* $d^2\vec{F}$ *exerted by the external air on the slice and the weight of the slice* $d^2\vec{P}$ *are such that* $d^2\vec{P} + d^2\vec{F} = \vec{0}$, *with* $d^2\vec{P} = -g\rho(z)\,dz\,dS\,\vec{u}$ *where ρ(z) is the density of air at that altitude.*

Is the statement above in italics true?

□ yes □ no □ dont' know

Explain your answer:

• *Question 4*

Is the following statement in italics true?

*The weight of a column of atmospheric air is equal to the action of this column over the ground.*

□ yes □ no □ dont' know

Explain your answer:

* It was restated orally that the collisions were elastic and that the container was fixed to a solid support.

The answer rates, which were very different depending on the statements made, show that the equivalence of these statements is far from clear.

***Table B1*** *- Answers from trainee teachers for each statement; the system concerned is in brackets. All the statements are accurate*

| N = 19 | Accurate | Non accurate | Don't know |
|---|---|---|---|
| Statement 1 (one molecule) | 5 | 10 | 4 |
| Statement 2 (molecules piled up) | 5 | 9 | 5 |
| Statement 3 (a slice of atmosphere) | 8 | 3 | 8 |
| Statement 4 (a column of atmosphere) | 11 | 5 | 3 |

Either because it was usual or because of the macroscopic nature of the system, statement 4 elicited a relatively high score of correct answers. The lowest number of right answers concerned the isolated molecule, being very unusual. Note that if statement 4 is presented in first place, then agreement can be unanimous as was the case during one survey carried out with a small group of 13 students in the third year of university.[157] The variability observed among trainee teachers is very probably due to the juxtaposition of ritual questions with others which were significantly more thought provoking.

In the case of the degree students, we were able to collect their opinions after they had seen the explanation in Appendix A. These students unanimously judged it to be "worth it, despite the extra teaching time involved" (half an hour). When questioned explicitly on this topic, 12 of the students said they found it enjoyable. Here are a few of their comments:

- "It made me understand just how difficult teaching is, and especially the importance of accurate explanations. Also, it gave me insight into the physical meaning, whereas before I had been looking mainly at the purely mathematical aspect of the question. We went further than simple questions about problems. I think that this is very important in teaching: it means that you can shape your course better, and especially, it allows you to take your distance and respond better to questions from the pupils."
- "It explains a fundamental question in simple terms which is not at all easy to explain intuitively to pupils."
- "There is a certain amount of difficulties in analysing the system reduced to a single molecule."
- "To deal with an exercise properly, it is always preferable to fully understand, both physically and mathematically, the problem in hand."
- A general comment: "I think it is very good to deal with each of the subjects at a more fundamental level; this helps you have clearer ideas and to be much more precise."
- "I like asking myself questions I have never asked before."

One comment from an experienced teacher in a one to one interview also merits attention:

(*A single molecule in a container*)

- "What that means is that being at the bottom of the container, and so... transmitting the weight, er, via the bottom of the container, or bouncing around inside the whole container exerting forces, it all comes to the same thing. Why does it all amount to the same thing? Huh?"

(*After explanation*)

- "Ah yes, normally when we study a gas we ignore gravity... we don't bring in a gravitational field (...). You showed in that container why the pressure was greater here (*lower face*) than there (*upper face*); that was shown with *g*."

157 Degree course "Physics for teaching", Université Paris Diderot, 2007.

# Appendix C

## CAUSAL LINEAR REASONING

As far as reasoning is concerned, one of the most common tendencies is to drastically reduce the analysis of functional dependencies, with one cause and one effect being the preferred scenario. Hence, in physics (we won't deal with other areas such as economics) we have: "If the density of a gas decreases, its pressure therefore decreases", "$U = IR$, if the resistance increases, the voltage $U$ increases". Such familiar statements say nothing about initially relevant variables such as the temperature of the gas or the current respectively. When the many variables are all taken into account, there is a particular form which gives structure to common reasoning: this is causal linear reasoning,[158] whose features have been characterised on the basis of work by S. FAUCONNET,[159] J.L. CLOSSET[160] et S. ROZIER.[161]

In short, common reasoning frequently adopts the structure of a linear chain of implications, in which each link is a single phenomenon ($\phi$), bearing on the evolution of a single quantity: $\phi_1 \rightarrow \phi_2 \rightarrow \phi_3 \rightarrow ...\phi_n$.[162]

Once again, we can observe the natural tendency of reasoning to simplify analysis by reducing the number of factors to be considered.

The following is an example of a common response,[163] in this particular case explaining the increase of pressure when a perfect gas is compressed adiabatically:

158 VIENNOT L. (2001) *Reasoning in Physics - The part of common sense*, Dordrecht: Kluwer Ac. Pub., Chapter 5.

159 FAUCONNET S. (1981) *Etude de résolution de problèmes : quelques problèmes de même structure en physique*, Thesis, Université Paris 7. And previous reference, 95-99.

160 CLOSSET J.L. (1983) *Le raisonnement séquentiel en électrocinétique*, Thesis, Université Paris 7, and reference note 158, 99-103. See also: SHIPSTONE D.M. (1984) A study of children's understanding of electricity in simple DC circuits, *European Journal of Science Education*, **6** (2) 185-198.

161 ROZIER S. (1988) *Le raisonnement linéaire causal en thermodynamique classique élémentaire.* Paris, thesis, Université Paris 7. Also: ROZIER S. & VIENNOT L. (1991) Students' reasoning in thermodynamics, *International Journal of Science Education*, **13** (2) 159-170. And reference note 158, 103-108.

162 See this work, Chapter 2.

163 Reference in note 161.

*volume decreases → number of particles per unit volume increases → number of collisions increases → pressure increases.*

As we can see, variations in temperature and the speed of the molecules are not considered. However, a more specific feature concerns the status of the arrow, which is used here to symbolize the relation between two successive phenomena in the explanatory sequence. The arrow interposed between predicates could represent a logical implication, *i.e.* "hence". Alternatively, the arrow could indicate the occurrence of a subsequent event, "next". Be that as it may, the overall impression is one of the comfortable ambiguity of the connector "then", failing as it does to disambiguate between a logical implication and a surreptitious chronological sequence (Table C1).

***Table C1*** *- Different languages, same ambiguity in meaning: the intermediate term evokes both a logical implication and a chronological sequence*

| Level ↓ | French | English | Spanish |
|---|---|---|---|
| Logical | donc | hence | por eso |
| Intermediate | alors | then | entonces |
| Chronological | ensuite | next | después |

This narrative version of the explanation itself arises from a number of different indicators, in particular the very frequent use of the future tense ("if the density increases, the pressure *will* increase") sometimes overtly found in the comments collected during the foundation surveys of this analysis. Thus:

For two springs in series hung from the ceiling, with the bottom end being pulled by the experimenter:[164]

"The first spring will extend, then *after a while*, the second will also extend."

Or:

To explain the expansion of an ideal gas during heating at constant pressure (ROZIER ibid.):

"*Heat is supplied → temperature rises → pressure increases → volume increases*", then, faced with the objection that the transformation is isobaric: "First of all, the piston is locked in place; heat makes the temperature and pressure rise; then the piston is free to slide, the volume increases, and the pressure returns to its previous value as *before*."

The arrows in the explanatory summary proposed above not only indicate a logical link, they also have a more or less explicit chronological connotation. This narrative form of reasoning has been observed among students and school pupils in a range of contexts, and with remarkable frequency. Although this scenario might be

**164** Already mentioned. Reference in note 159.

appropriate for analysing a series of effectively sequential events, it is completely inappropriate for modelling the evolution of multivariate systems under quasi-static or quasi-steady conditions; that's to say, when we assume that several variables, each characterising the whole system, evolve (quasi) simultaneously under the steady constraint of several simple relations expressing a state of equilibrium or a steady regime. Typically, $pV = nRT$ (in the usual notation) for a perfect gas, or the balance of energy flows for a bolometer. The terms in italics contrast term by term with those which characterise causal linear reasoning (Table C2).

- the envisaged events are often described simply, using a *single variable*;
- more or less explicitly, these events are understood to be *successive*,
- and hence as temporary, or at least they are *temporarily considered*.

What is lost in this narrative type of explanation is the elementary approach of phenomena where the internal propagation times of the system being studied are neglected by comparison with those which characterise overall evolution of the system. Naturally, this means that simultaneity and steady behaviour are sacrificed.

***Table C2*** *- Contrasting terms between causal linear reasoning and analyses currently employed in physics.*

| **In quasi-static physics** | **An example** | **Linear causal stories** |
|---|---|---|
| – several variables | 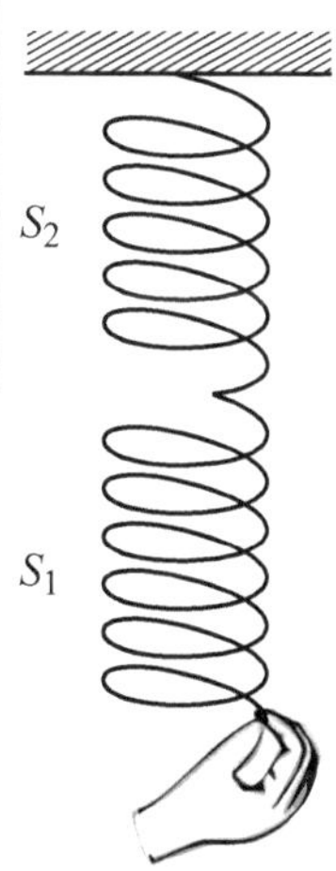 | – simple phenomena (one variable each) |
| – change (almost) simultaneously | | – are taken into account sequentially (hence) |
| – constrained by permanent relationships | | – temporarily |
| $\vec{F}_{ext}(t) = \vec{T_1}(\text{same } t) = \vec{T_2}(\text{same } t)$<br>$\Delta l_T = \Delta l_1 (\text{same } t) + \Delta l_2 (\text{same } t)$<br>$F_{ext}$: Force exerted by an experimenter on the lower end; $T_1$, $T_2$: tensions of each spring; $\Delta l_1$, $\Delta l_2$: extensions of each spring, $\Delta l_T$ total extension. | | *A symptomatic comment:*<br>"The first spring ($S_1$) will extend then, after a while, the second ($S_2$) will also extend." |

# Appendix D

## When physics should conform to beliefs: pierced bottles[165]

The remarkable thing about the problem presented here is its great age, and the historical frequency of its associated errors, despite the long-established clarifications in the literature. This is in a sense a case of cultural obstinacy,[166] the error in this case.

For those with some grasp of physics it is quite clear that the pressure in a fluid increases with depth. It might be better to add "fluid at equilibrium" for the statement to be correct, but if omitted, it should be implicit in the context. For authors writing for the general public, it is always tempting to *demonstrate* this dependence on depth, and therein lies the probable origin of the recurrent illustrations of the type shown in Figures D1 (a). The range of the jets of water emerging from a pierced bottle is shown as increasing with the distance between hole and free surface. That suggests (some would say, demonstrates) that at each hole the pressure increases with depth.

Except that, if we actually do the experiment (it's preferable to use an overflow system to maintain a constant water height) we observe the situation shown in Figure D1 (b): a maximum range for the mid-height hole, and equal ranges for holes symmetric about the mid-hole.

How is it that so many authors, from Leonardo da Vinci onwards, have been able to use this experimental setup without physically and systematically putting it to the test? Hardly for want of refutation. Torricelli had already cleared up the situation around 1640,[167] and since then there has been no shortage of objections. Despite

165 For more details see: Planinšič G., Ucke C. & Viennot L. (2011) Holes in a bottle filled with water: which water jet has the largest range? Muse project of the EPS-PED (www.eps.org, select *Education* and then select *MUSE*). See also Planinšič G. & Viennot L. (2011) Jets and inverted jets: a matter of differences 2011 (www.eps.org, select *Education* and then select *MUSE*).

166 Josip Slisko has long been developing this point of view, on the basis of a sample of articles including all those mentioned here (except French authors): Slisko J. (2009) Repeated errors in physics textbooks: what do they say about the culture of teaching? *Physics Community and Cooperation*, GIREP 2009, University of Leicester. See also Slisko J. (2006) Errores en los libros de texto de física: ¿cuáles son y por qué persisten tanto tiempo? *Sinectica*, **27**, 13-23.

167 Voir Slisko J. (2009) refs. in previous note.

all this, still recently (in France) three books[168] make use of this example, where it would seem that the physics has been reinvented just for the convenience of instructive visualisation.

**(a)**

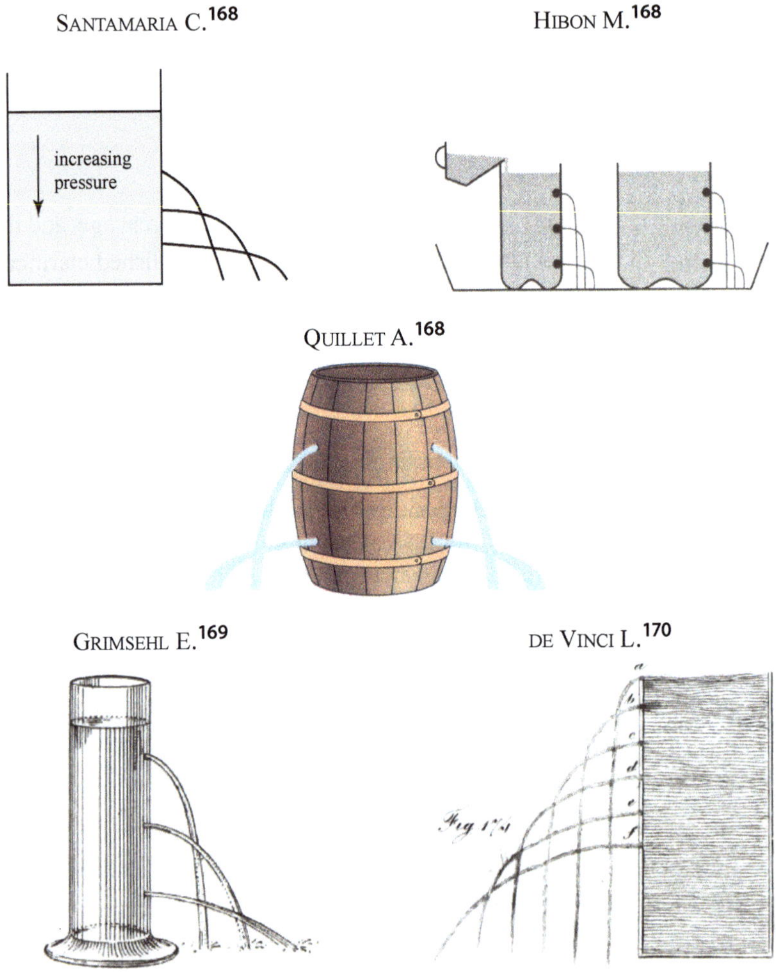


168 Hibon M. (1996) *La physique est un jeu d'enfant*, Activities in scientific wakening, Armand Colin, Paris, p. 126 ; Santamaria C. (2007) *La physique tout simplement*, Ellipses, Paris, p. 7 ; Quillet (1993) *Teach-yourself encyclopaedia*, Quillet, Paris, p. 220. Concerning examples from other countries, see numerous references in (Slisko 2006, 2009: note 164) and (Atkins 1988: note 169).

169 Grimsehl E. (1912) Lehrbuch der Physik, Leipzig und Berlin: Verlag von B. G. Teubner, figure 261, p. 239 cited by Slisko J. (2006) Errores en los libros de texto de física: ¿cuáles son y por qué persisten tanto tiempo? *Sinectica*, **27** (13-23) p. 19.

170 Leonardo da Vinci (1828) *Del moto e misura dell' acqua di Leonardo da Vinci. A spese di Francesco Cardinali*, Bologna. Digitised copy from Harvard College Library, Google books (http://www.archive.org/stream/raccoltadautorii10card#page/n537/mode/2up).

**(b)**

ATKINS J.K. [171] PLANINŠIČ G.

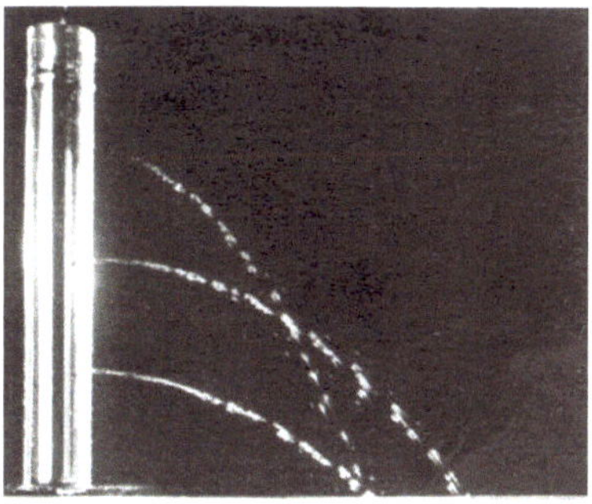

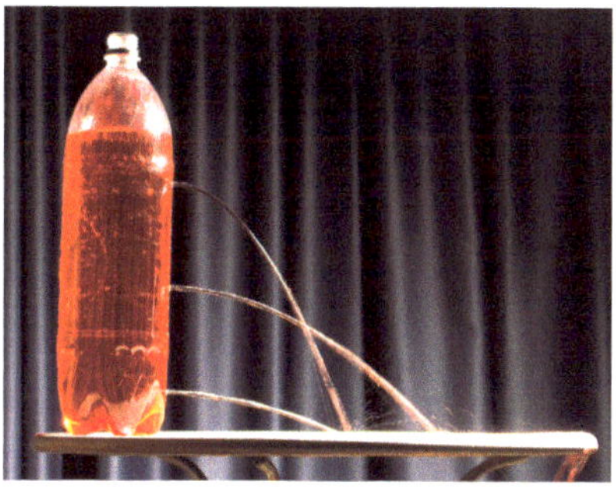

***Figure D1*** *- The range of jets of water from a pierced bottle*

(a) *Again and again we find illustrations indicating that the lower jet has the greatest range. Patently, this would show that hydrostatic pressure increases with depth: a far from recent ritual.*

(b) *Actually the speed at which the water emerges does indeed increase with the distance from the hole to the water surface, but the time taken to fall increases with the distance of this hole from the table (non-linear factors). Ultimately it's the product of these two distances which determines the range of the jets. If the values of these distances are interchanged (holes symmetric with respect to the medium) then the ranges are equal.*

Clearly, the problem here is, on the one hand, a lack of vigilance but also the confusion between the exactitude of a conclusion (pressure increases...) and the value of a demonstration. The very principle of this demonstration is contentious (see the calculations below): an attempt is being made to show the increase in hydrostatic pressure even though it is a dynamic situation. The standard formal treatment of the situation, using the so-called "BERNOULLI's theorem",[172] involves calculating the exit speed as if the water had fallen from the surface in free fall, at constant pressure – a supreme irony. It is difficult, therefore, to relate the range of the jet to an exit pressure, so we go for another variable: the exit speed of the water, which is horizontal. This is dependent on (the square) of the depth of the hole.

However, here we encounter another difficulty: this is not the only variable that counts. The time taken for the water to fall outside the bottle also contributes to the horizontal advance of the water before impact. Ultimately, it is the product of these two factors which explains the observation. Exit speed and time taken to fall are linked (via their square) respectively to the distance from the hole to the surface, and

171 ATKINS J.K. (1988) The great water-jet scandal, *Physics Education,* IOP Publishing, **23**, 137-138.

172 When the flow of a perfect (*i.e.* inviscid) *incompressible* fluid (of density $\rho$), is *steady* in the presence of gravity g, the following quantity is *conserved along a streamline*: $v^2/2 + p/\rho + g\,z$ where $v$ is the flow speed and $p$ is the pressure at the same height $z$. For irrotational flow, this quantity is the *same throughout the fluid.*

to its distance to the table. The product of these distances correlates well with the effective impact locations.

One factor instead of two: very often that's all it takes for a "simple" explanation to become absurd. Absurd indeed: imagine a hole right at the level of the table on which the bottle stands; would a jet emerging from this point reach the maximum range?

To be sure, it is a little more complicated to provide the explanatory elements here, but we can avoid inconsistency by seeking a process of reasoning which at least some non-specialist readers will find accessible. We also avoid the ridiculous situation of being in blatant contradiction with an experiment anyone can reproduce for themselves by using a hot nail to make holes in a plastic bottle.

*Standard calculation*

The classic approach is to use BERNOULLI's theorem for this situation. This presupposes that the flow regime is steady and that the liquid is incompressible and inviscid. The application of this theorem to two points on a streamline (Fig. D2) located respectively on the free surface ($A$) and at the water outlet ($B$, at height $h$) gives:

$$v_A^2/2 + p_A/\rho_{\text{water}} + g\,z_A = v_h^2/2 + p_B/\rho_{\text{water}} + g\,z_B \quad (1)$$

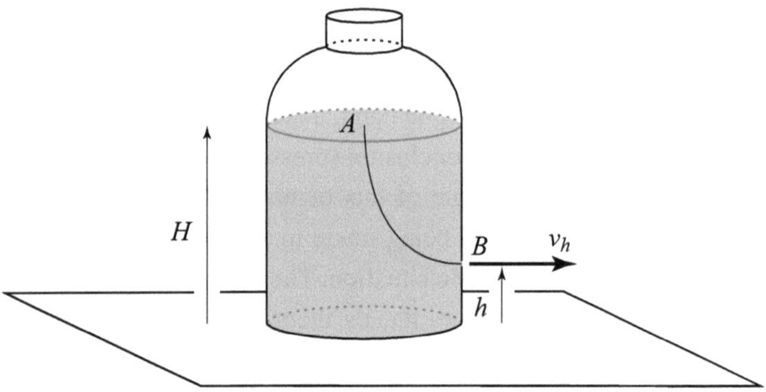


***Figure D2*** *- Application of* BERNOULLI*'s theorem at two points on a streamline for the flow of water within the bottle.*

We then assume that the flow speed $v_A$ at the free surface (whose area is much greater than that of the hole) is virtually zero compared with the exit flow speed $v_h$ from the hole being considered. The water pressure is the same at the free surface as at the exit hole, *i.e.* atmospheric. Equation (1) is then written

$$p_{\text{atm}}/\rho_{\text{water}} + g\,z_A = v_h^2/2 + p_{\text{atm}}/\rho_{\text{water}} + g\,z_B$$

Therefore, the square of the speed, $v_h^2$, is proportional to the difference $z_A - z_B$, *i.e.* the difference between the height of the water surface, $H$ and that of the hole, $h$:

$$v^2_h = 2g\,(H - h) \quad (2)$$

It is interesting to note that this formula is the same as that obtained with the free fall model. The relationship $p_h - p_{\text{atm}} = \rho g(H - h)$, where $p_h - p_{\text{atm}}$ is the difference in pressure between the altitudes h and H, cannot be used because it assumes hydrostatic conditions which clearly do not hold in the case of an accelerating fluid stream.

A second factor affects the range of the water-jet: the time $t_{ff}$, of the water's free fall, such that $h = \frac{1}{2}\, g\, t_{ff}^2$.

The square of this time, $t_{ff}^2$, is proportional to the altitude $h$ of the hole as measured from the table onto which the jet falls:

$$t_{ff}^2 = 2h / g \qquad (3)$$

Assuming the water jet is horizontal at the exit hole, its range is obtained by multiplying the velocity $v_h$, at the exit hole by the time of free fall, $t_{ff}$.

$$d = v_h\, t_{ff} = 2[h(H - h)]^{1/2} \qquad (4)$$

The range of the jet therefore depends on the product of the two distances whose sum is $H$. With a stable level of water in the bottle, a steady flow regime is observed where the maximum range is obtained for the mid-height hole:

$$h_m = H / 2$$

Two holes symmetric about $h_m$ will, in this model, give rise to jets of the same range.

It should be made clear that practical realisation of this experiment confirms the predicted sequence in the observed water jet ranges, but not their numerical values. Questions remain about the holes as well as about the viscosity of the water. For further details, see the online article cited with the title of this appendix (Planinšič *et al.* 2011).

# Appendix E

## Reactions of trainee journalists and scientific writers confronted with inconsistency

(*Appendix written with Stéphanie Mathé*)

*The benefit of consistency in physical theories: the opinions of non-specialists*

The idea of consulting third-year university students who are not intending to follow careers as teachers, physicists or engineers on this topic might seem curious. However, the authors of the study[173] considered it could be an interesting line of approach. It is widely argued that the pleasure derived from following a line of reasoning is one of those luxuries that is the preserve of the specialist. Emphasising the consistency and economy of physical theories and their effectiveness in predicting phenomena might be a preoccupation justified solely by future professional needs and requiring considerable expertise. The apprentice journalists and writers mentioned here are (for this study) non-specialists, and moreover somewhat mixed in terms of their previous training.[174]

We wished to duplicate with these students a previous investigation we had carried out with physics students (second or third year, with the results briefly stated in Chapter 6) based on the "instructional hot-air balloon" which had been modelled in a highly contentious manner. Those students had been asked and given support to analyse an exercise in which it was claimed that the internal and external pressures of the air were equal. In Chapter 6, the reader will find a critique of such an assumption, which implies a zero resultant for the forces exerted by the air (internal and external) on each part of the envelope, and using a more global approach, favours no particular direction for the action of the air; it is hard to see how any vertical thrust could result from this. The suggested clarification summarised in Chapter 6 (Fig. 6.1) relies on the central theoretical plank of hydrostatics, namely the existence of pressure gradients. This makes the lack of vigilance of the teachers who set the problem all the more remarkable. It is no less striking to see the reactions of students for whom half

173 Mathé S. & Viennot L. (2009), ref. in note 149.

174 See caption in Table E1.

an hour's clarification, individually or in groups, was sufficient. To be sure, in this first study, as in that referred to in this appendix, it was not the conceptual performances of the subjects which were finally assessed, but their reactions to the value of the experience they had just had. We find many comments testifying to intellectual satisfaction, and also several indications of retrospective annoyance: "How come that's the first time anyone told me that?".

The first question that arises when investigating the answers of non-specialists is: will we observe, among students for whom physics itself has been of limited importance both in the past and in the future, the same type of reaction, *i.e.* associating the emotional with the conceptual? If so, the assumption that consistent reasoning only appeals to the elite does not stand.

This study could cast light on the training of future journalists or scientific writers, and, in fact, pleads for such training to be specifically oriented. We shall return in the conclusion to this sensitive and rather polemical issue.

*Main results of the survey*

We surveyed 14 students enrolled on a course in preparation for professions in scientific mediation. They were interviewed on the subject of a simulated article concerning the operation of a hot-air balloon (Box E1), where the usual assumption as to the equality of internal and external pressures is shown.

### *How a hot-air balloon works*

A little physics...

How does a large object like a hot-air balloon manage to rise into the air? According to ARCHIMEDES' principle! We already know how this works with bodies immersed in water. Well, it works in the same way in air, as in any other fluid. The hot-air balloon, in the same way as anybody immersed in air, undergoes a vertical upward force equal to the weight of the fluid displaced, *i.e.* the surrounding air.

For the balloon to rise, ARCHIMEDES' upthrust has to overcome two other forces tending to pull it downwards: namely, the weight of the air inside the envelope, and the weight of the basket plus contents.

On the day of takeoff, let's assume this weight to be fixed. As for ARCHIMEDES' upthrust, it depends on both the density of the air displaced and the volume of the body immersed in this fluid, *i.e.* the volume of the balloon. During the flight however, these two parameters are also fixed.

So it's the weight of the air inside the envelope that can be modified, and this is what can tip the balance. But how? By means of the burners placed at the opening of the envelope, which will heat this air. The density of a gas depends on its pressure and temperature. The pressure is the same inside and outside, because of the opening at the bottom of the envelope, through which the air can communicate. As for temperature, warming the air makes it less dense and therefore less heavy.

From a certain temperature (in practice around 100 °C) onwards the weight of the air in the envelope will be low enough for ARCHIMEDES' upthrust to prevail over the weight of the basket plus that of the internal air. The balloon can then rise!

***Box E1** - A simulated article given to students.*

The structure of the interview is shown in Box E2.

### Step 1

a) Ask for the student's opinion concerning the fact that the pressures inside and outside are said to be the same.

b) Try to invalidate this assumption using the two reasons presented in the text – one in terms of symmetry, and the other using a local mechanistic argument.

c) Explain ARCHIMEDES' principle to bring out the idea that pressure depends on altitude ($z$) and density ($\rho$).

d) Draw a graph of the external and internal pressures versus the altitude and stress that there is a difference in pressure, in particular at the top of the envelope.

e) Link this difference in pressure to the upward force resulting from the action of the air inside and outside the envelope; this is more easily done with the model of a cubic balloon.

### Step 2

Here we tried to compile the students' value judgements about this more rigorous approach:

a) Ask if everything has been well understood.

b) Ask if they feel able to explain it all over again to someone else.

c) Ask what kinds of public could be presented with this problem.

### Step3

a) Ask if they valued this type of discussion and if they derived pleasure from the whole interview. Ask them to rate this pleasure on a scale from 1 (low) to 4 (very high).

b) Similarly, ask them to evaluate a kind of 'interest': 'cost in time' ratio on a scale of 1 (low) to 4 (very high).

c) Ask if, as journalists, they would consider it worth introducing diagrams, graphs, formulae, etc. to explain the functioning of the hot-air balloon as in the example, or if they think it would overburden the reader to no useful purpose.

***Box E2** - The different steps of the interview.*

The results summarised in Table E1 bring out the following features.

First reactions do not focus on the physics of the phenomenon but, as might be expected in a training course of this sort, on how readable the article is: this is hardly surprising.

On questioning, analysis of the contentious assumption was laborious. Many students were irritated when they realised that they had a feeble grasp of the subject. Rather than take up the argument, they referred back to the article of which they seemed to be at pains to take a distance. Of the two arguments available to counter this assumption, associated with spherical symmetry or zero force over the entire envelope respectively, the second appears by far the best to trigger a reaction.

***Int.*** (*the interviewer*): *Let's see what happens on a small part of the envelope. If the pressures are the same inside and outside*, the forces of pressure are equal so…

***Céline***: They'll cancel each other out.

***Int.***: Yes, they are the same on both sides, OK? Everywhere on the envelope, you repeat the same thing, so what happens to the entire envelope if they cancel each other out?

***Céline***: Well it… Ah yes… But you told me it wasn't a party balloon… Well, it doesn't move… I don't know… It doesn't move, does it?

***Int.***: There we are. If they cancel each other out, there's no overall force applied to it. So isn't what he says rather strange? And don't forget the basket below, and its weight.

While not enough to bring the discussion to an end, this reaction can be strong. Often it is only after several steps of this type that the brusque distrust vis à vis of the text becomes apparent (see Table E1 for the paths followed for this type of shift). This happens in particular when it is really understood, not to say rediscovered, that the Archimedean thrust involves a pressure gradient (Table E1).

Comments then become very critical:

***Nuno***: There (*in the article*), he means that the pressure is the same… So there's something wrong here! He's only talking about the temperature and the pressures would be the same… But in that case, ARCHIMEDES' upthrust wouldn't apply, would it?

***Laurence***: So in that case, there's no pressure gradient according to what he's saying, so you have no idea where his upthrust comes from!

***Int.***: So at the top of the balloon…

***Carine***: You have the inner pressure, which is greater than the outer.

***Int.***: Exactly.

***Carine***: That makes it rise. That's not at all what it says here! (*She shows the article*)

***Int.***: No, that's not really explained in the article.

***Carine***: Oh no! Because honestly, when I read it, I didn't understand it this way. The fact that he says that the density depends on the pressure and the temperature… To me, it really was at a constant pressure all the time and by making the temperature vary… But in fact, the temperature makes the pressure vary too. That's what isn't explained. ■

***Table E1** - Steps in students' intellectual paths: Awareness of the inconsistency and a critical attitude.*

| Name and "scientific origin"<br>α: architecture<br>β: biology<br>μ: mathematics<br>ι: technology<br>φ: physics | From the start | First oral question about the assumption | Argument of symmetry<br><br>Step 1b | Local explanation<br><br>Step 1b | Origin of ARCHIMEDES' upthrust and link with the pressure gradient<br><br>Step 1c | Plotting the graph<br><br>Steps 1d and 1e | When asked if they felt able to explain<br><br>Step 2 |
|---|---|---|---|---|---|---|---|
| Nuno (β) | $C_0$ | A / C | | A | A | A | |
| Ludovic (β) | $C_0$ | A | A | | | A | C |
| Laurence (β) | $C_0$ | A | | A | A / C | A | |
| Carine (β) | | | A | | | A / C | |
| Adeline (φ) | | | A | | | A / C | |
| Céline (β) | | | | A / C | | | |
| Côme (ι) | | | | A | | | C |
| Damien (β) | | | | A | | | C |
| Dima (β) | | | | A | | A / C | |
| Anna (μ) | | | | A | | A / C | |
| Marion (φ) | $C_0$ | | | A | | A | C |
| Emmanuelle (α) | $C_0$ | | | | A | A / C | |
| Laura (φ) | | | | | A | A | C |
| Thomas (φ) | | | | | | A | C |

'A' indicates when the students clearly showed their awareness of the inconsistency.

'$C_0$' indicates some signs of a critical attitude from the start, not yet focused on the assumption.

'C' indicates when the students first used their awareness of the inconsistency to criticize the article or to retrospectively criticize their own attitude during the interview.

There is nothing in common between the comments at this stage and those at the beginning. It is no longer a question of desperately trying to recall unreliable school memories, since the reasoning process is under way and the tools used are recognised as useful. The resulting satisfaction is patently clear (13 out of 14 students give a mark of 3 or more in a scale from 1 to 4). The following extracts from the interview are testament to this:

***Côme***: Ah yes! I always like understanding things! You've just made me a happy man! I like it!

***Laurence***: Yes. I'd say 3. As it began with something that wasn't very clear in my mind, it's all the more pleasant to discover things when you are working with someone else and you are actively engaged. The satisfaction is even greater because of all the difficulties you overcome.

***Nuno***: I even think that... It's not only useful, it's also... I've understood, I've reasoned, I've made my little graph in order to be able to understand all the chains of reasoning, with this formula you explained to me, so yes! For me, it's absolutely a must!

***Int.***: What rating would you give?

***Nuno***: I would say 4. The more you learn, the better it is.

***Int.***: Ah! You give a 4 now?!

***Nuno***: Well... Let's say 3... It's difficult to choose a mark between 3 and 4! Let's say 3.5! Because we needn't overdo it! But what you did here was essential.

***Marion***: If going deeper into the explanation means that we have to go through this, then I think it's important.

***Int.***: Do you think it's worth using this explanation, if you ever had to write an article, even though it takes more time, even though it's more complicated?

***Laurence***: If it's worth it? That's an odd question! Yes, of course it's worth it!

***Int.***: You could have said this article was sufficient...

***Laurence***: Well, I'm not in a position to say 'no, it's useless!'

***Int.***: You could have said it was too complicated...

***Laurence***: No no! It's not complicated at all! Honestly, I'm really bad at physics, what I say is based on what I learnt at secondary school and it's not my thing! So, yes it's worth it!

There are a certain number of remaining comments as to what effect the experience might have on them for their future professional life. At that moment in the interview, after they felt they had acquired some understanding and/or something had been added to their reasoning potential, the students professed intentions that some may judge to be naïve. It is these intentions which are emphasised here, not their prophetic value. It seems that these non-specialists are not averse to the challenge of certain intellectual expectations. The idea of a necessary compromise with editorial realism is there, but for most of the students, this did not imply giving up this intellectual rigour.

***Int.***: What kind of readers do you feel you would be able to explain this to?

***Laurence***: To teenagers. Because I think you start to learn that kind of thing in secondary school and it would be pretty difficult if I couldn't use any formulae. I would have a lot of trouble explaining it to younger people, I think.

This survey provides indices which support the following line of argument: *if* the purpose of training future journalists includes giving them an idea of how models and theories operate, even if only in elementary physics, the public in question would not offer the strong resistance we so often hear about.

### *In conclusion*

The ideas supported by this limited investigation are therefore on two levels.

First of all, that of the supposed elitism which would underly the project, making the students think a little more than with our teaching rituals. We cannot see the relevance of such a point of view in these results. Our first experimental subjects (Chap. 6) were not preselected, while those in this survey were preselected on the basis of their probably reduced competence. Most of the students instantly expressed interest and pleasure. We hesitate to say this was an unwarranted attitude so as not to depreciate it.

The second level of discussion raised by these results concerns the objectives of the training. Certain reactions to the study described here appear to be opposed to the idea that consistency in a physical description has anything to do with the training of a journalist. The underlying view of science itself is involved, of course. Talking of the *broad degree* of internal consistency and the predictive power of physics can still upset those who might worry about an unsubtle scientism. Furthermore, popularisation is not the work of a journalist or teacher; each specialism draws on highly specific skills. Hence, evidently, if this study claimed to have brought the discussions related to these points of view to a conclusion, it would be far short of the target.

At least the study supports the idea that it would be useful to make a clear definition of the scientific training objectives for future professionals in journalism and/or "mediators in Science". It would undoubtedly be worth having second thoughts before asserting that there is no place for the illustration of internal consistency and the predictive power of the theories under discussion.

# Appendix F

## "Facilitating elements" of communication: Year 11 students ranking the risks of misunderstanding

*Written with Ivan* Feller*, drawing on his thesis:*

*Using non-school documents in the Year 11 classroom: School-oriented reduction, reactions to implausibility, and outcomes of an action research (In French: Usage scolaire de documents d'origine non scolaire - Eléments pour un état des lieux et étude d'impact d'un accompagnement ciblé en classe de seconde)*

http://tel.archives-ouvertes.fr/tel-00366318/fr/ and note 60

It is commonly advocated that students should be educated to become responsible citizens. Improving their critical sense is one of the main arguments used. In the perspective of life-long education, they should be put in contact with scientific media, which they should learn how to "read", in the broad sense. How not to subscribe to such wishes? All the more so that, in addition, motivation is expected to be a side benefit.

There are numerous studies available concerning museum visits. By contrast, using non-academic resources in the classroom is a less well documented field: how can a teacher in his/her class fruitfully use written scientific documents intended for a general public, given the constraints of the syllabus and the limitations on time? An investigation at Year 11 level provides some insights on how such a challenge can be faced.

First of all we need to redefine the objective in rather less general terms than those of "developing a critical sense".

One possible option, that adopted by the study in question, is to work on documents (paper, for practical reasons) which are related to the concepts being taught in that particular year, though not necessarily restricted to them. One of the goals is to spread the main message, without being restricted to merely what is in the programme. Another desirable objective is that the student should learn to detect any implausible details the author may have introduced. We can give a name to these simplifications, metaphors, exaggerations or other improbable elements which a document's author may sometimes use when trying to make himself understood: "facilitating elements". One activity suggested by the study[175] is that of guiding the students towards a comparison of the probable consequences of these facilitating elements. We shall illustrate these two themes using the previously mentioned ET document, in which a couple of aliens were observing the Earth from a very long way off.

**The scene takes place in the year 2000.**

We could wonder whether the alien can answer the question. In fact, if the alien knew when mankind dicovered fire, he would be able to deduce that Earth is 3000,000 light-years from their planet (that is about $3.10^{18}$ km). It shows that the propagation of light is not instantaneous.

***Figure F1*** *- The ET document*

The main message can be stated as followed: "*Given the finite value of the speed of light, a message carried out by light reaches us with a time delay.*"

Reducing this message to its purely scholastic level (at Year 11) consists of focussing on just one thing here: we can calculate the distance between planets if we know how far in the past the event was observed, *i.e.* the era in which the hairy earthlings were *then* making fire. In short, by using the values of two variables we can calculate the third with the formula $d = vt$. The following part of the message would then be wasted: what we are seeing happened a long time ago, since, because of the finite speed of light, the received signal has taken time to reach us. For those who unders-

175 Here we find a central preoccupation in the study as described in the previous appendix: training the students to take a position on the elements of a text which are liable to pose a problem of consistency.

tand, these two statements might seem to be equivalent. The fact is, they are not expressed with the same frequency before, and after, training.

In the category of "facilitating elements" we can say that these aliens speak our language, that the two of them are on some minuscule planet, and that their telescope is capable of picking out human hair on Earth, etc. More seriously, *i.e.* with more probable consequences for comprehension, we observe that the Earth is shown as being much too big for its presumed distance (of the order of a few hundred thousand light-years). Clearly, we can understand that it is easier to show the continents so that Earth is recognisable. However, in so far as the orders of magnitude pertaining to the main message are concerned, this veers towards inconsistency. A critical reading of this document (or so one hopes, at least) by the students should lead them to this incipient analysis

The study relies on the idea that comprehension of the main message and the ordering of the associated "facilitating elements" are actually linked objectives. The students' progress would develop with both of them.

On this basis, an experiment in Year 11, repeated three years in a row, involved a total of 94 students. Each year it took up only four and a half hours of teaching time based on the three documents illustrated, including this one by way of example. Without going into the details of the experiment, for each of the first two documents the students were encouraged to give their opinion on the main message, the "facilitating elements", and on how interesting they found the document. On each of these points, a discussion with half the class was led by the teacher on a selection of replies from the same students to a prior individual questionnaire. On a number of occasions–at the start and end of each session, when working on their own on the third document, and at the end of the year–a questionnaire provided an update on the state of how each student's thought processes were progressing.

The results were very stable from one year to the next, allowing the corresponding participants to be grouped together.

In the case of the "ET" document, and before any teaching on the topic had started, more than half the students followed the school text: that is to say, with two pieces of data, we can calculate the distance between planets. After the corresponding session of instruction, this proportion of students dropped to one quarter, a sign both of progress and of a persistent difficulty. An extract from the discussion provides a good illustration of the type of work involved:

| Transcript | Our comments |
| --- | --- |
| The teacher (T) reads some students' previous responses to a questionnaire: | |
| T– What do you think of response 5): "**The aim of this document is to get us realise that a light-year is a very large distance**"? | *The teacher launches the discussion about the role of the distance between the planets in the main message.* |
| *A– Well no, because it's to understand that the light does not arrive immediately.* | |
| T– So what do you think of the student's comment? Is it false? | |
| *B– This is not the goal, it's true but it's not the main goal.* | *The teacher provides students with an opportunity to make the distinction: "correct from the point of view of physics / identification of the main message".* |
| T– So, what is the main goal then? | |
| *C– Well, it's to get us understand that the light takes a certain time to reach us ...* | |
| *D– In fact, it's response 2) that is correct.* | |
| T– Number 2)?: **The author is trying to get us understand that the arrival of light is not immediate, because the extraterrestrial who lives in year 2000 sees an image of prehistoric Earth.** What do you think of that? | *The teacher seeks to check whether the group agrees with this idea.* |
| *All– Yes!* | *Collective agreement.* |

***Box F1*** *- An excerpt from the guided discussion on the main message of Doc ET.; T: teacher, A,B,C,D: students. Bold fonts indicate students' responses to a questionnaire (previously selected).*

As for the "facilitating elements" in the document, it is their total absence from the students' initial responses before the teaching process had started which is significant. After the corresponding session, a third of the students mentioned the size of the Earth as being at the top of the list of contentious parts of the document: here again, this rather modest progress provides a measure of the difficulty. In the meantime, the intervening discussion gave a more positive image.

| Transcript | Our comments |
|---|---|
| T– **There are only two inhabitants on the ET's planet.** What do you think of this response? | *The teacher points to an implausible element (FE).* |
| *A– It's true, but it's a drawing, we are not going to represent all the people on Mars.* | *This student considers it as anecdotal.* |
| T– All right. The question is to know if it hinders the comprehension of the main message. | *The teacher raises the question as to whether this is a potential obstacle (PO)?* |
| *All– Well no, it doesn't matter!* | *It is not considered by the students as consequential for comprehension.* |
| (…) | *(Same type of dialogue about responses 3 and 5: the teacher lets the students talk freely. The spontaneous reproduction of the process is a clue to the students' progress.)* |
| T– Yes, all right. Wait, I'll read: **It is not possible to see these details (the Earthlings are hairy and they are trying to make fire) with a telescope, especially if we're 300,000 light-years away.** So, what do you think of this? Can this blur the main message? | |
| *D– Well no! Scale isn't the point for the author. A big or small telescope, it doesn't make any difference.* | *Spontaneously, the student turns to the problem of scales. These students seem to consider this FE as harmless.* |
| T– All right. | |
| *E– Well, look at response 6). He says that the scale is incorrect. But it is in no way the purpose of the author to have appropriate scales.* | *This is the case for two other students (E, C).* |
| *C– Yes, we aren't going to draw one planet and then the other one thousands of light-years away; there isn't enough room.* | *The student considers the illustrator's point of view. But the potential obstacle associated with the inappropriate scaling is not identified.* |
| *D– In fact, this way of showing us things helps us understand.* | |
| T– All right, so, I ask you: what do you think of the student who said: **we get the impression that the ETs are close to the Earth?** | *The teacher tries to orientate the discussion towards the representation of the Earth.* |
| *C– Well yes, it's the only planet that we can see with the naked eye.* | |
| T– Well then, if we are so close to the Earth, is the light going to take 300,000 years before reaching us? | |
| *D– Well no, it will take much less time.* | |

| Transcription du débat | Commentaires |
|---|---|
| T– Then let us pose the question the other way round: if we had represented the Earth at a distance of 300,000 light-years, what would have we seen? | *Gradually, the teacher stresses the inconsistency linked to the size of the Earth.* |
| *All– Well, nothing!* | |
| *F– A point.* | |
| T– Then why did the author represent the Earth like that? | |
| *G– Well it's so we'll understand that it's the Earth.* | *The student underlines the reason why the drawing is like that: in order to facilitate the identification of the Earth in the image.* |
| T– All right, so if we draw the Earth as a point, can we understand the main message all the same? | |
| *H– Of course we can, it is what they say that enables us to understand: I see Earthlings ...* | *This reason is soon challenged.* |
| T– All right. So, was it necessary to represent the Earth like that? | |
| *All– No!* | |
| T– You told me first of all: the scale isn't a serious problem. What do you think of this? | |
| *I– Well yes, there, it does matter.* | |

***Box F2*** *- Excerpt from the discussion on "facilitating elements" from the ET document. T: teacher; A, B, C, D: students. Bold fonts indicate students' responses to a questionnaire (previously selected)*

The second document used, which is more complex both in its structure and in the concepts involved, appeared more appropriate for group work than for individual work. We again see the students' difficulties in going beyond the purely scholarly reading for understanding the main message, and their failure to spot the contentious "facilitating elements" in the document.

While they were better equipped to deal with the third document, these difficulties nevertheless underline how hard it is for students to benefit from reading such texts.

Is this then about going against the stream, swapping a formal chore for the carefree and rewarding reading of an illustrated document? Listen to the participants at the end of their third and last session, this time done on their own:

– "I found this work very instructive; it enabled me to study a document in depth and to write a critique of a text written for the general public. I think that a document of this type should contain simple but clear explanations. When some notions are simplified too much, they sometimes end up being barely understandable. (…)

Doing this work, made me realise that the goal of physics was to extract relevant pieces of information from a document and to select the main ones, and not just to make calculations or carry out experiments."

- "I found this work rather difficult because developing criticism based on arguments involves a lot of thought. It gave me an opportunity to do something new by approaching this subject in a literary way, and I spent several hours on this."
- "In the first place, this work helped me to develop my critical sense. Now, I've got a better understanding as to why it is necessary to consult several sources to be sure about the different information that we talked about. And, it is also interesting to analyse a document, and to know exactly what we understand and what we do not understand".
- "This work convinced me that it is necessary to take time to read an article, and not just glance through it, because it is interesting. We should not give up because of something we do not understand. It also made me understand that I could be interested in a subject which, at first sight, would not have attracted me if I had not been obliged to work on it (...). After all, this work made me much more aware of things I would not have thought of otherwise."

One point is worth noting. At the end of the training period, and in relation to what is described in this appendix, a survey over the whole year and its activities revealed a better reaction among students who did not envisage scientific specialisation as compared with those of their classmates aiming for *Terminale S* (Year 13 science). The findings of this survey indicate that future scientists more often showed themselves to be sceptical of the utility of such work. This observation is yet to be confirmed and, if need be, taken into account, so the usual calls for a critical sense can be accompanied by better-informed actions.

# Bibliography


Allende J.E. (2008) Academies Active in Education, *Science*, 321.

Ambrose B.S., Shaffer P.S., Steinberg R.N. & McDermott L.C. (1999) An investigation of student understanding of single-slit diffraction and double-slit interference, *American Journal of Physics* **67** (2) 146-155.

Andersson B. (1989) The experiential Gestalt of Causation: a common core to pupils preconceptions in science, *European Journal of Science Education*, **8** (2) 155-171.

Artigue M., Menigaux J. & Viennot L. (1990) Some aspects of students' conceptions and difficulties about differentials, *European Journal of Physics*, **11**, 262-267.

Aspect A., Balian R., Balibar S., Brezin E., Cabane B., Fauve S., Kaplan D., Léna P., Poirier J.-P., Prost J. (2004) *Demain la physique*, Odile Jacob, Paris.

Atkins J.K. (1988) The great water-jet scandal, *Physics Education,* IOP Publishing, **23**, 137-138.

Bachelard G. (2002) *The Formation of the Scientific Mind* (Translation Mary McAllester-Jones) Clynamen Press, Manchester.

Bennhold C. & Feldmann G. (2005) Instructor Notes On Conceptual Test Questions, *In Giancoli Physics - Principle with applications*, 6th Edition, Pearson, Prentice Hall, 290-291.

Besson U. & Viennot L. (2004) Using models at mesoscopic scale in teaching physics: two experimental interventions on solid friction and fluid statics, *International Journal of Science Education*, **26** (9) 1083-1110.

Besson U., Borghi L., De Ambrosis A. & Mascheretti P. (2007) How to teach friction: Experiments and models. *American Journal of Physics*, **75** (12) 1106-1113. (Video available at http://fisica.unipv.it/didattica/Energia/ENG/irrevers.htm)

Boohan R. & Ogborn J. (1997) Differences, energy and change: a simple approach through pictures, *New ways of teaching physics*, Proceedings of the Girep International Conference 1996 in Ljubliana, Oblack S., Hribar M., Luchner K., Munih M., Board of Education, Slovenia.

Chabay R.W., & Sherwood B.A. (2002) *Matter & Interactions II: Electric & Magnetic Interactions*, John Wiley & Sons, New York.

Chauvet F. (1996) Teaching colour: designing and evaluation of a sequence, *European Journal of Teacher Education*, **19**, n°2, 119-134 (http://www.lar.univ-paris- diderot.fr/sttis_p7/color_sequence/page_mere_fr.htm).

Chauvet F. (2004) Une simulation pour explorer un modèle cinétique de gaz en seconde, *Bulletin de l'Union des Physiciens*, **98**, n°866, 1091-1105 (online training documents: http://www.epi.asso.fr/revue/articles/a0306d/Gaz_a.htm).

Chavannes I. (1907) *Leçons de Marie Curie aux enfants de nos amis* [Marie Curie's lessons to our friends' children], EDP Sciences, Paris (2003).

Colin P. & Viennot L. (2001) Using two models in optics: Students' difficulties and suggestions for teaching, *Physics Education Research, American Journal of Physics Sup.*, **69** (7) S36-S44.

COLIN P., CHAUVET F. & VIENNOT L. (2002) Reading images in optics: students' difficulties and teachers' views, *International Journal of Science Education*, **24**, 3, 313-332.

COLIN P. (2011) Enseignement et vulgarisation scientifique : une frontière en cours d'effacement ? Une étude de cas autour de l'effet de serre, *Spirale*, 48, 63-84.

CRAWFORD F.S. (1965) *Berkeley Physics Course Vol. 3*: *Waves*, McGraw-Hill Company, New York.

CURIE M. & CHAVANNES I. (1907) *Physique élémentaire pour les enfants de nos amis*, Work coordinated by LECLERCQ B. (2003), EDP Sciences, Paris.

DE BROGLIE L. (1941) *Continu et discontinu en physique quantique*, Albin Michel, Paris.

DEVAUD M. & TREINER J. (2010) Théorie cinétique de la pesée d'un gaz, *Bulletin de l'UDPPC*, **104** (928) 1021-1024.

DIU B. (2000) *Traité de physique à l'usage des profanes*, Odile Jacob, Paris.

DIU B., GUTHMANN C., LEDERER D. & ROULET B. (1989) *Mécanique Statistique*, Hermann, Paris, 350-352.

DUIT R. (2006) *Students' and Teachers' Conceptions and Science Education* (http://www.ipn.uni-kiel.de/aktuell/stcse/stcse.html).

EINSTEIN A. (1905) Zur Elektrodynamik bewegter Körper, *Ann. d.Ph.*, **17**, 892-921 (translation SOLOVINE, GAUTHIER-VILLARS, 1955, **5**). See also: EINSTEIN A. (1907) Relativitätsprinzip und die aus demselben gezogenen Folgerungen, *Jahrbuch der Radioaktivität*, **4**, 411-462 & **5**, 98-99.

FELLER I. (2008) *Usage scolaire de documents d'origine non scolaire en sciences physiques. Eléments pour un état des lieux et étude d'impact d'un accompagnement ciblé en classe de seconde*, Doctoral thesis, Université Paris Diderot (http://tel.archives-ouvertes.fr/tel-00366318/fr/).

FELLER I., COLIN P. & VIENNOT L. (2009) Critical analysis of popularisation documents in the physics classroom. An action-research in grade 10, *Problems of education in the 21st century* (PEC) **17**, 72-96.

FEYNMANN R. (1965) *The character of Physical Law*, Massachussetts Institute of Technology, the M.I.T. Press, Cambridge.

GALILI I., GOLDBERG F. & BENDALL S. (1991) Some reflections on plane mirrors and images, *The Physics Teacher*, **29** (7) 471-477.

GALILI I. & HAZAN A. (2000) 'Learners' knowledge in optics: interpretation, structure, and analysis', *International Journal in Science Education*, **22** (1) 57-88.

GOLDBERG F. & MCDERMOTT L.C. (1986) Student Difficulty in Understanding Image Formation by a Plane Mirror, *The Physics Teacher*, **24** (8) 472-480.

GUNSTONE R. & WATTS R. (1985) Force and Motion. In DRIVER R., GUESNE E. & TIBERGHIEN A., *Children's ideas in science*, Milton Keynes: Open University Press.

HUNT R. & MILLAR R. (2000) *Science for Public Understanding, London: Heinemann, 21st century science* (jb56@York.ac.uk).

JACOBI D. (1987) *Textes et images de la vulgarisation scientifique*, Peter Lang, Berne.

JEANNERET Y. (1994) *Ecrire la science*, PUF, Paris.

JURDANT B. (1975) La vulgarisation scientifique, *La Recherche*, **53**, 149.

KAHNEMAN D. (2012) *Thinking Fast and Slow*, Penguin books, London.

Kautz C.H., Heron P.R.L., Loverude M.E. & McDermott L.C. (2005) Student understanding of the ideal gas law, Part I: A macroscopic perspective, *American Journal of Physics*, **73** (11) 1055-1063.

Kautz C.H., Heron P.R.L., Shaffer P.S. & McDermott L.C. (2005) Student understanding of the ideal gas law, Part II: A microscopic perspective, *American Journal of Physics*, **73** (11) 1064-1071.

Klaassen K. & Linjse P. (2010) Interpreting students' discourse and teachers' discourse in science classes: an underestimated problem?" In *Designing Theory Based Teaching-Learning sequences for science Education*, Kortland K. & Klaassen K. (Eds.), CDβ Press, Utrecht.

Kress G. & van Leeuwen T. (1996) *Reading Images: the Grammar of Visual design*, Open University Press, Utrecht.

Laukenmann M., Bleicher M., Fuller S., Gläser-Zikuda M., Mayring P. & Rhöneck C.V. (2003) An investigation of the influence of emotional factors on learning in physics instruction, *International Journal of Science Education*, **25** (4) 489-507.

Leroy-Bury J.L. & Viennot L. (2003) Doppler et Römer : physique et mathématique à l'œuvre, *Bulletin de l'Union des Physiciens*, **859**, 1595-1611.

Lévy-Leblond J.M. (1986) *Mettre la science en culture*, Anais, Nice.

Mathé S. & Viennot L. (2009) Stressing the coherence of physics: Students journalists' and science mediators' reactions, *Problems of education in the 21st century* (PEC) **11**, 104-128.

Maurines L. (1992) Spontaneous reasoning on the propagation of visible mechanical signals, *International Journal of Science Education*, **14** (3) 279-293.

McDermott L.C. (1996) *Physics by Inquiry*, Vol. 1 and II, John Wiley & Sons, New York.

Mervis J. (2013) Transformation is Possible if University Really Cares, *Science*, **340** (6130), 292-296.

Millar R. (2006) Twenty First Century Science: Insights from the design and implementation of a scientific literacy approach in school science, *International Journal of Science Education*, **28** (13) 1499-1521.

Ogborn J., Kress G., Martins I. & McGillicuddy K. (1996) *Explaining Science in the Classroom*, Open University Press, Buckingham.

Ogborn J. (1997) Constructivist metaphors of learning science, *Science & Education*, **6**, 121-133.

Ogborn J. (2004) Physics education. In J. Ogborn (Ed.) *Physics now*, ICPE-IUPAP (http://web.phys.ksu.edu/icpe/Publications).

Ogborn J. (2010) Science and Commonsense. In M. Vicentini and E. Sassi (Eds.): *Physics Education: recent developments in the interaction between research and teaching*, Angus and Grapher Publishers, New Dehli.

Pintrich P.R., Marx R.W. & Boyle R.A. (1993) Beyond cold conceptual change: The role of motivational beliefs and classroom contextual factors in the process of conceptual change, *Review of Educational Research*, **63** (2) 167-199.

Planinšič G. (2004) Color light mixer for every student, *The Physics Teacher*, **42**, 138-142.

Planinšič G. & Viennot L. (2010) Shadows: stories of light, Muse project of the EPS-PED (www.eps.org, select *Education* and then select *MUSE*).

Planinšič G. & Viennot L. (2011) Jets and inverted jets: a matter of differences Muse project of the EPS-PED (www.eps.org, select *Education* and then select *MUSE*).

PLANINŠIČ G., UCKE C. & VIENNOT L. (2011) Holes in a bottle filled with water: which water jet has the largest range? Muse project of the EPS-PED (www.eps.org, select *Education* and then select *MUSE*).

PSILLOS D. (1995) Adapting Instruction to Students' Reasoning. In D. PSILLOS (Ed.), "*European Research in Science Education*", Proceedings of the second PhD Summerschool, Leptokaria, Thessaloniki: Art of Text, 57-71.

RHÖNECK C.V., GROB K., SCHNAITMANN G.W. & VÖLKER B. (1998) Learning in basic electricity: how do motivation, cognitive and classroom climate factors influence achievement in physics?, *International Journal of Science Education,* **20** (5) 551-565.

RIGAUT M. & VIENNOT L. (2002) Réduire le théorème du centre d'inertie : jusqu'où ? *Bulletin de l'Union des Physiciens,* **841**, 419-426.

ROCARD Y. (2007) *Science Education Now*, Report EU22-845, European Commission, Brussels (http://ec.europa.eu/research/science-society/document_library/pdf_06/report-rocard-on-science-education_en.pdf).

ROQUEPLO P. (1974) *Le partage du savoir*, Le Seuil, Paris.

ROZIER S., VIENNOT L. (1991) Students' reasoning in thermodynamics, *International Journal of Science Education,* **13,** n°2, 159-170.

SALTIEL E. & MALGRANGE J.L. (1980) Spontaneous ways of reasoning in elementary kinematics, *European Journal of Physics*, Vol. 1, 73-80.

SHIPSTONE D.M. (1984) A study of children's understanding of electricity in simple DC circuits, *European Journal of Science Education,* **6** (2) 185-198.

SLISKO J. (2006) Errores en los libros de texto de física: ¿cuáles son y por qué persisten tanto tiempo? *Sinectica,* **27**, 13-23.

SLISKO J. (2009) Repeated errors in physics textbooks: what do they say about the culture of teaching? *Physics Community and Cooperation*, GIREP 2009, University of Leicester.

VIENNOT L. (1982a) L'implicite en physique : les étudiants et les constantes, *European Journal of Physics*, **3**, 174-180.

VIENNOT L. (1982b) L'action et la réaction sont-elles bien égales et opposées ? *Bulletin de l'Union des physiciens*, **640**, 479-485.

VIENNOT L. & RAINSON S. (1999) Design and evaluation of a research-based teaching sequence: The superposition of electrics fields. *International Journal of Science Education,* Special issue: *Conceptual Development in Science Education* (continued), **21** (1) 1-16.

VIENNOT L. (2001) *Reasoning in Physics The part of common sense*, Kluwer Ac. Pub., Dordrecht.

VIENNOT L. (2003) *Teaching physics*, Kluwer Ac. Pub., Dordrecht.

VIENNOT L. (2004a) The design of teaching sequences in physics - Can research inform practice? Lines of attention. Optics and solid friction, In *Research on Physics Education, Proceedings of the International School of Physics Enrico Fermi (Italian Society of Physics), Course CLVI*, Societa Italiana di Fisica, Bologna, 505-520.

VIENNOT L. (2004b) The design of teaching sequences in physics. Can research inform practice? DOPPLER and RÖMER. In E. F. REDISH & M. VICENTINI (Eds. Research on Physics Education. Course CLVI, SIF Varenna. Amsterdam: IOS press, 521-532.

VIENNOT L. & LEROY J.L. (2004) DOPPLER and RÖMER: what do they have in common?, *Physics Education*, **39**, issue 3, 273-280.

VIENNOT L. (2005) La transmission des valeurs de la science, *Science et Avenir* Special edition, n°144, p. 35.

VIENNOT L., CHAUVET F., COLIN P. & REBMANN G. (2005) Designing Strategies and Tools for Teacher Training, the Role of Critical Details. Examples in Optics, *Science Education*, **89** (1) 13-27.

VIENNOT L. (2006a) Modélisation dimensionnellement réductrice et traitement "particulaire" dans l'enseignement de la physique, *Didaskalia*, **28**, 9-32.

VIENNOT L. (2006b) Teaching rituals and students' intellectual satisfaction, *Phys. Educ.*, **41,** 400-408 (http://stacks.iop.org/0031-9120/41/400).

VIENNOT L. & KAMINSKI W. (2006) Can we evaluate the impact of a critical detail? The role of a type of diagram in understanding optical imaging, *International Journal of Science Education*, **28** (15) 1867-1895.

VIENNOT L. (2007) La physique dans la culture scientifique : entre raisonnement, récit et rituels, *Aster*, special No. « Science et récit », **44**, 23-40.

VIENNOT L. & PLANINŠIČ G. (2009) *The siphon: a staging focused on a systemic analysis*, Muse project of the EPS-PED (www.eps.org, select *Education* and then select *MUSE*).

VIENNOT L. (2010) Physics education research and inquiry-based teaching: a question of didactical consistency. In K. Kortland (Ed.) *Designing Theory-Based Teaching-Learning Sequences for Science Education*, Cdβ press, Utrecht, 39-56.

VIENNOT L., PLANINŠIČ G., SASSI E. & UCKE C. (2010) *Various experiments in fluid statics*, Muse project of the EPS-PED (www.eps.org, select *Education* and then select *MUSE*).

VIENNOT L. (2011a) Le poids de l'air, le choc des molécules : il fallait bien que BOLTZMANN s'en mêle, *Bulletin de l'UDPPC*, **105** (932) 313-315.

VIENNOT L. (2011b) *Floating between two liquids*, Muse project of the EPS-PED (www.eps.org, select *Education* and then select *MUSE*).

VOKOS S., SHAFFER P.S., AMBROSE B.S., MCDERMOTT L.C. (2000) Student understanding of the wave nature of matter: Diffraction and interference of particles, *American Journal of Physics* **68**, S42-S51.

WELTIN H. (1961) A paradox, *American Journal of Physics*, **29** (10) 712-711.

WHITE R. & GUNSTONE R. (1992) *Probing Understanding*, Falmer Press, London.

WITTMANN M.C., REDISH E.F. & STEINBERG R.N. (2003) Understanding and Addressing Student Reasoning about Sound, *International Journal of Science Education*, **25** (8) 991-1013.

WOSILAIT K., Heron P.R.L., SHAFFER P.S., MCDERMOTT L.C. (1999) Addressing student difficulties in applying a wave model to the interference and diffraction of light, *American Journal of Physic,s* **67**, S5-15.

MIX
Papier aus verantwortungsvollen Quellen
Paper from responsible sources
FSC® C105338

If you have any concerns about our products,
you can contact us on
**ProductSafety@springernature.com**

In case Publisher is established outside the EU,
the EU authorized representative is:
**Springer Nature Customer Service Center GmbH**
**Europaplatz 3, 69115 Heidelberg, Germany**

Printed by Libri Plureos GmbH
in Hamburg, Germany